The
Reproductive
System

The Reproductive System

Titles in the Understanding the Human Body series include:

Understanding
THE HUMAN BODY

The Reproductive System

Pam Walker and Elaine Wood

LUCENT BOOKS®

THOMSON

GALE

San Diego • Detroit • New York • San Francisco • Cleveland • New Haven, Conn. • Waterville, Maine • London • Munich

LIBRARY OF CONGRESS CATALOGING-IN-PUBLICATION DATA

Walker, Pam, 1958–
    The reproductive system / by Pam Walker and Elaine Wood.
      v. cm. — (Understanding the human body)
Includes bibliographical references and index.
Contents: Male reproductive tract—Female reproductive organs—Fertilization
and development—Diseases of the reproductive system—Medical technologies
of the reproductive system.
    ISBN 1-59018-152-2
    1. Human reproduction—Juvenile literature.   [1. Reproduction.]
I. Wood, Elaine, 1950– II. Title. III. Series.
    QP251.5 .W35 2003
    612.6—dc21

                                                2002011841

Printed in the United States of America

CONTENTS

FOREWORD

Since Earth first formed, countless creatures have come and gone. Dinosaurs and other types of land and sea animals all fell prey to climatic shifts, food shortages, and myriad other environmental factors. However, one species—human beings—survived throughout tens of thousands of years of evolution, adjusting to changes in climate and moving when food was scarce. The primary reason human beings were able to do this is that they possess a complex and adaptable brain and body.

The human body is comprised of organs, tissue, and bone that work independently and together to sustain life. Although it is both remarkable and unique, the human body shares features with other living organisms: the need to eat, breathe, and eliminate waste; the need to reproduce and eventually die.

Human beings, however, have many characteristics that other living creatures do not. The adaptable brain is responsible for these characteristics. Human beings, for example, have excellent memories; they can recall events that took place twenty, thirty, even fifty years earlier. Human beings also possess a high level of intelligence. Their unique capacity to invent, create, and innovate has led to discoveries and inventions such as vaccines, automobiles, and computers. And the human brain allows people to feel and respond to a variety of emotions. No other creature on Earth has such a broad range of abilities.

Although the human brain physically resembles a large, soft walnut, its capabilities seem limitless. The brain controls the body's movement, enabling humans to sprint, jog, walk, and crawl. It controls the body's internal functions, allowing people to breathe and maintain a heartbeat without effort. And it controls a person's creative talent, giving him or her the ability to write novels, paint masterpieces, or compose music.

Like a computer, the brain runs a network of body systems that keep human beings alive. The nervous system relays the brain's messages to the rest of the body. The respiratory system draws in life-sustaining oxygen and expels carbon dioxide waste. The circulatory system carries that oxygen to the body's vital organs. The reproductive system allows humans to continue their species and flourish as the dominant creatures on the planet. The digestive system takes in vital nutrients and converts them into the energy the body needs to grow. And the immune system protects the body from disease and foreign objects. When all of these systems work properly, the result is an intricate, extraordinary living machine.

Even when some of the systems are not working properly, the human body can often adapt. Healthy people have two kidneys, but, if necessary, they can live with just one. Doctors can remove a defective liver, heart, lung, or pancreas and replace it with a working one from another body. And a person blinded by an accident, disease, or birth defect can live a perfectly normal life by developing other senses to make up for the loss of sight.

The human body adapts to countless external factors as well. It sweats to cool off, adjusts the level of oxygen it needs at high altitudes, and derives nutritional value from a wide variety of foods, making do with what is available in a given region.

Only under tremendous duress does the human body cease to function. Extreme fluctuations in temperature, an invasion by hardy germs, or severe physical damage can halt normal bodily functions and cause death. Yet, even in such circumstances, the body continues to try to repair itself. The body of a diabetic, for example, will take in extra liquid and try to expel excess glucose through the urine. And a body exposed to extremely low temperatures will shiver in an effort to generate its own heat.

Lucent's Understanding the Human Body series explores different systems of the human body. Each volume describes the parts of a given body system and how they work both individually and collectively. Unique characteristics, malfunctions, and cutting edge medical procedures and technologies are also discussed. Photographs, diagrams, and glossaries enhance the text, and annotated bibliographies provide readers with opportunities for further discussion and research.

A Broad Perspective on Reproduction

What is the most important group of organs in a species? Many have argued that it is the reproductive system. If a species could not reproduce, it would not exist after the last surviving member of the first generation had died. If a group of living things could not reproduce, the other body systems would be of limited value because when enough critical parts had worn out or become injured, individual members of the group would die, one by one, until the species became extinct.

The first living things on Earth were simple, one-celled organisms. These pioneers of life developed a one-parent style of reproduction that is still being used today by similar creatures. However, as more complex, multi-cellular organisms evolved, a two-parent method of reproduction developed with them.

The One-Parent Method

Every species of living thing reproduces. However, the manner of reproduction is not the same for all of them. In the process called cell division or fission, one-celled creatures divide to create two new cells, both of which are identical to the parent. Cell division is said to be an asexual method because, with only one parent, reproduction is accomplished without the participation of a sexual partner. Budding is another type of asexual reproduction. Simple organisms, like sponges, form buds or extensions from the parent. These buds eventually separate from the parent, forming a carbon copy of the original.

Fission in one-celled creatures seems like a fairly simple task: The parent cell splits apart to create two daughter cells just like itself. However, the apparent simplicity of cell division is an illusion. Before division can occur, the parent cell's genetic material must be redistributed so that both new cells will end up with a complete set of heritable information identical to that of the parent. The offspring produced by fission and other asexual processes are clones, carbon copies of the parent.

The heritable information within a cell is vitally important because it is both the blueprint and the instruction book for that cell. Most cells carry this information, their genetic material, in chromosomes, long strands of deoxyribonucleic acid (DNA). Just before cell division these strands engage in a complicated dance to ensure that

After distributing its genetic material into two sets of heritable information, a paramecium undergoes asexual reproduction.

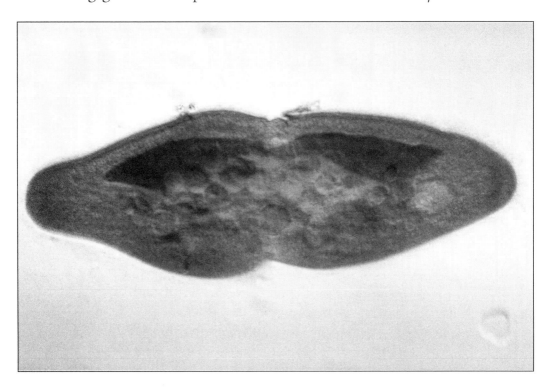

each of the two resulting cells receive its fair share of genetic material.

The Two-Parent Method

Most of the complex, multicelled organisms, such as dogs, trees, and humans, do not reproduce asexually. They use a more elaborate system called sexual reproduction. In sexual reproduction, genetic material from two parents combines to form offspring. As a result, the offspring is similar to both parents, but not exactly like either one. In this way, sexual reproduction creates one-of-a-kind individuals, all of which have slightly different traits.

It is well known among scientists that the species, or groups of individuals, that are best at staying alive are those that produce offspring with a variety of traits. Variety among individuals within a species helps the entire species survive. To understand the advantages of variety, consider a hypothetical species of mammals living in an area that has a warm climate year-round. If two parents were to produce a group of offspring that are identical to one another, they would all have the same size ears, the same type of fur, the same voice, and the same intellect. In the event of a drastic and unexpected change in the environment, like an Ice Age, the parents and all of their identical offspring would perish because they lacked characteristics enabling them to survive in cold temperatures.

However, if two parents in this pretend species produced a group of offspring in which each individual had some unique traits, there might be one or two in the group that had extra thick coats of fur. These individuals might have a better chance than others in their family of surviving the colder climate. The survivors, in turn, would have the opportunity to mate and produce offspring of their own.

Sexual reproduction creates variability among offspring. The process of sexual reproduction is more complex than fission. Most species that carry out sexual reproduction have two distinctly different types of individuals: males and females. Each type of individual produces specialized cells called gametes that are assigned the job of ensuring the continuation of the species. It is the union of these cells that creates a new, unique individual.

The Genetic Governor

The genetic material that governs the actions of cells in complex organisms is contained in threadlike structures called chromosomes. Different species of organisms have different numbers of chromosomes. Humans have a set of forty-six chromosomes, and most cells in the human body contain forty-six chromosomes. There is one exception: The gametes, or sex cells, have been modified to contain only half a set of chromosomes.

Reproduction starts when two gametes fuse, forming a zygote. Since each human gamete possesses twenty-three chromosomes, a human zygote packs a complete set of forty-six chromosomes. Shortly after formation, a zygote begins dividing, making new cells. Each of the resulting cells also contains forty-six chromosomes. This type of cell division, called mitosis, is essential for growth and is similar to fission in one-celled organisms. During mitosis, a parent cell produces two new cells with a set of genetic material just like its own. The commonly used term "divide" does not adequately describe the complexity of the mitotic event.

The Dance of Mitosis Begins

In mitosis, the DNA or genetic material within a cell's chromosomes is copied and then distributed evenly between the two new cells. This entire process must be carried out very precisely because any change in DNA results in a change in the genetic material. Some of these changes, or mutations, are good but many cause damage to cells.

Scientists describe mitosis as a series of steps or stages called interphase, prophase, metaphase, telophase, and

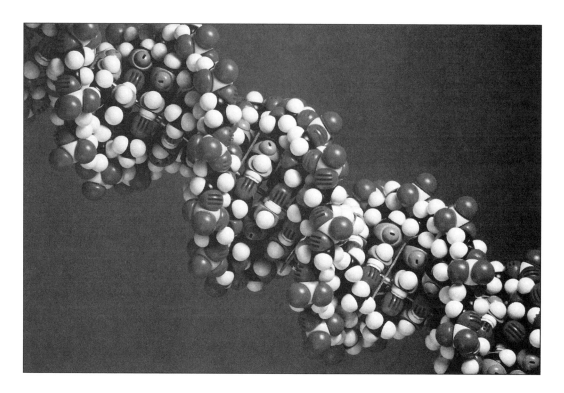

A scientific model illustrates a single strand of DNA.

cytokinesis. Most of a cell's life is spent in interphase. It is during interphase that the cell grows and performs the jobs it was designed to do. In interphase, a cell's DNA exists as chromosomes within the cell's nucleus, a large structure held together by a membrane. When a cell is ready to undergo mitosis, each chromosome replicates, or makes an exact copy of itself. The original strand of chromosome remains attached to its new copy by a filament called a centromere. The two identical strands, joined together at their center by a centromere, are called chromatids. Meanwhile, also undergoing replication are other essential structures in the cell's cytoplasm: mitochondria, ribosomes, and centrioles.

Once chromosomes have replicated, interphase is over and cell division then moves along at a rapid pace. In the next stage, prophase, the paired chromatids shorten and thicken into rodlike structures, better suited to moving around the cell without getting tangled. The membrane around the nucleus breaks down, releasing the chromatids into the cellular cytoplasm. At the same time, centrioles, structures that aid in cell division, move to opposite sides of the cells like boxers squaring off for a fight. Unlike boxing, however, mitosis is a cooperative event. Spindle fibers form between the two centrioles and work together like ropes and pulleys to move the chromatids around the cell.

In the next stage of mitosis, metaphase, spindle fibers move chromatids into the center of the cell. In this position, the sets of chromatids, still joined by their centromeres, are like paired dancers lined up in the center of a dance floor. One strand of each chromatid pair is attached to a spindle fiber that is controlled by centrioles located near the ends of the cell.

In the fourth stage, anaphase, the spindle fibers begin tugging on the chromatids, rupturing the centromeres and causing the chromatid pairs to separate. The middle of the dance floor is emptied as the tugging motion of the centrioles draws the once paired partners away from each other.

The Dance of Mitosis Concludes

The dance draws to a close in telophase, the point at which the chromatids reach the centrioles at the opposite sides of the cell. Now individual chromatids are renamed chromosomes, and events of mitosis begin to occur in reverse, like a movie played backward. The thick rods that make up each chromosome unwind and form long, thin fibers. A nuclear envelope forms around each forty-six-chromosome set, and the spindle fibers disappear. The elongated cell now possesses two complete nuclei.

Division of the cytoplasm, cytokinesis, is the last stage of mitosis. The cell membrane begins to pinch inward between the two nuclei. As it does, cytoplasm and other cellular structures are distributed evenly between the newly forming cells. Eventually, the two budding cells separate, each surrounded by their own cell membrane. Except for being smaller, each new cell is an exact replica

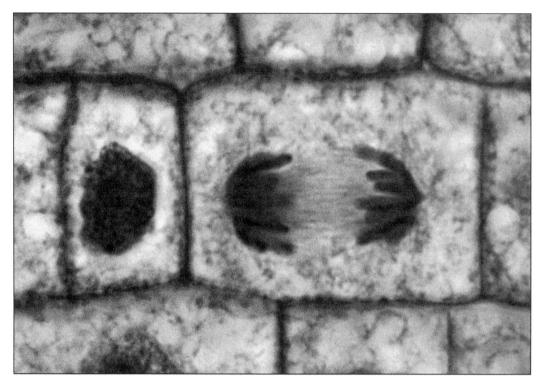

Chromatid pairs pull away from each other during the anaphase stage of an allium root's mitosis.

of the original cell. With time, the new cells increase the quantity of their cytoplasm so that they are exactly like their parent.

Mitosis is a sophisticated and exact system for producing new body cells, which are in constant demand as an organism grows and repairs damage. However, in multicellular organisms like humans, mitosis cannot supply the kind of cells needed to propagate the species. Reproduction requires a particular type of cell division, one that yields gametes.

The Dance of the Gametes Begins

The foundation of sexual reproduction is the union of gametes from two parents to form a zygote. For the zygote to have the correct total chromosome number, each of the gametes can only have half the number of chromosomes found in either parent. Therefore, cells produced by mitosis cannot act as gametes.

Gametes are specialized cells that carry just half the total number of chromosomes. Their only job is to combine with a gamete produced by a member of the opposite sex. Gametes are created through a distinctive type of cell division called meiosis. Meiosis is similar to mitosis in some ways. However, it is a more complex process because it has more stages and two sequences of cellular division. It is often separated into two phases, meiosis I and meiosis II.

When meiosis begins, the parent cell that will ultimately produce gametes has a full set of chromosomes. In interphase I, the parent cell makes two exact copies of its DNA, just like a cell preparing for mitosis. Therefore, each chromosome copies itself. The resulting, identical chromatids are linked by a centromere.

In prophase I, paired chromatids condense into rodlike structures and the nuclear membrane breaks down, as in mitosis. However, an event now occurs that does not happen in mitosis. Paired chromatids couple with another set of chromatids of the same size and shape. Then these double sets of paired chromatids swap some of their DNA in a process called crossing over. The purpose of this DNA exchange is to increase the genetic diversity found in each gamete. Meanwhile, centrioles move to opposite ends of the cytoplasm and spindle fibers form between them.

In metaphase I, spindle fibers align the double sets of paired chromatids in the middle of the cell. As in mitosis, each set of chromatids is attached to a spindle fiber that leads to centrioles on opposite ends of the cell.

Spindle fibers shorten in anaphase I, pulling the double pairs apart. One set of paired chromatids travels to one centriole while the other set is moved to the opposite centriole.

In telophase I, chromatids reach the centrioles and the cytoplasm divides to form two separate cells. However, the process of meiosis is not over. The paired chromatids must still be separated into individual chromosomes. Therefore, another division occurs in each new cell.

The Dance of the Gametes Concludes

Meiosis II is just like division in mitosis except for one important point: Chromatids do not replicate before prophase. In prophase II, centrioles again move to opposite sides of the cells and spindle fibers form between them, herding the paired chromatids into the middle of the cell. In this phase, in metaphase II, a spindle fiber is attached to each one of the joined chromatids.

In anaphase II, spindle fibers pull the paired chromatids apart, reeling one chromatid toward each centriole. At this point, the individual chromatids are called chromosomes again. In telephase II, nuclear membranes form around the two separate sets of chromosomes. Cytokinesis splits the cytoplasm in half in each of the cells, resulting in a total of four cells. Each of these has one-half as many chromosomes as the original cell. These cells are called gametes.

Spindle fibers of a lily in metaphase I of meiosis align double sets of paired chromatids.

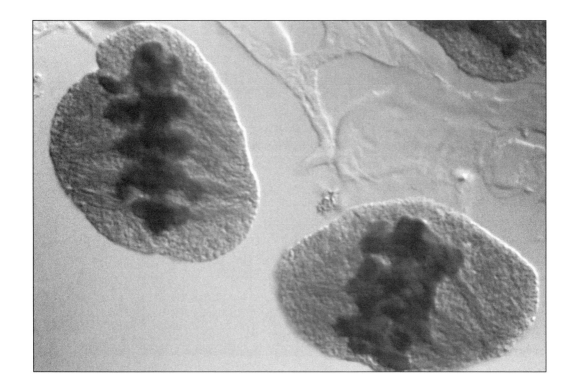

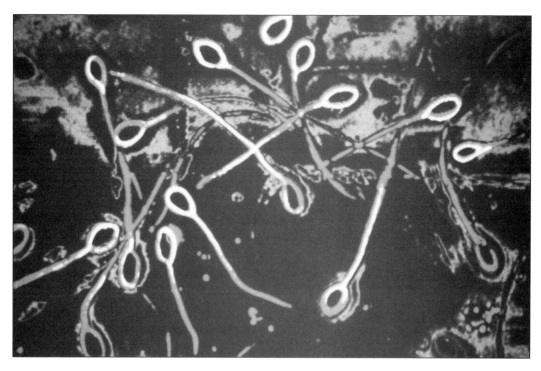

Gametes produced by males are called sperm and those by females are eggs. Even though meiosis is basically the same in each sex, the structures and processes involved differ. The male and female reproductive systems are highly specialized not only for making gametes, but also for enabling them to encounter gametes produced by the opposite sex.

Sperm cells carry a male organism's genetic material.

The Chromosome Wrap

The ability of a species to produce offspring ensures its survival. Living things reproduce in one of two basic fashions, either asexually or sexually. Simple, one-celled organisms use the asexual method, whereas most multicellular creatures have gone the road of sexual reproduction.

Some of the most successful organisms in the world have developed a method of sexual reproduction because it provides a species with variety among its offspring. If a group of living things has variety, it has a better chance of surviving a dramatic change in its environment.

Sexual reproduction requires the fusion of DNA from chromosomes of two parents. This DNA is ferried by sperm and egg, cells that are specialized gametes. Gametes result from a type of cell division called meiosis in which the normal chromosome number is reduced by one-half. When the two gametes fuse, each with their one-half load of precious cargo, the resulting zygote has a full set of DNA.

2 The Male Reproductive Tract

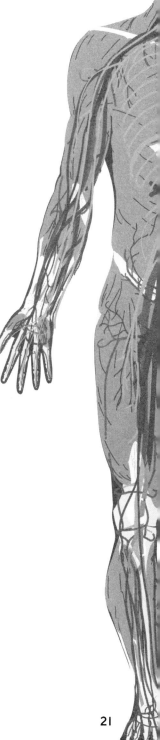

Most of the organs and organ systems in human males and females are identical. While it is true that many organs are larger in men than they are in women, hearts and brains, bones and stomachs are basically the same in both sexes. Male and female reproductive systems, however, differ greatly and these differences make it possible for the human species to continue.

In both males and females, the reproductive system has a unique status. It depends completely on all other parts of the body to support it. It is controlled by nerves, nourished by blood, regulated by hormones, oxygenated by the lungs, fed by digestive organs, protected by immune responses, and held in place by bones, muscles, and tendons. Yet, it does not directly support any other system. The nervous system could operate quite well without the reproductive organs, as could the digestive, circulatory, respiratory, immune, endocrine, and muscular/skeletal systems.

Nevertheless, it could be argued that the reproductive system is the most important group of organs in any species, for without it, the other body systems would not be needed.

Meeting the Male

The male reproductive system is composed of both external and internal organs and structures. The penis,

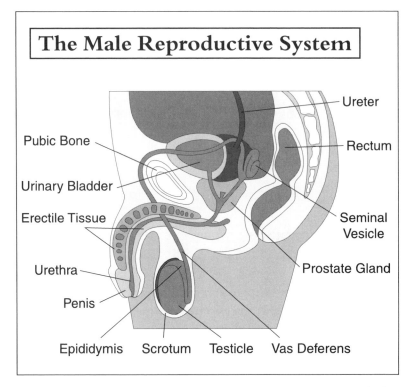

The Male Reproductive System

Ureter

Pubic Bone

Rectum

Urinary Bladder

Erectile Tissue

Seminal Vesicle

Urethra

Prostate Gland

Penis

Epididymis   Scrotum   Testicle   Vas Deferens

scrotum, and testes are located on the outside of the male's body. The vas deferens, urethra, prostate gland, and seminal vesicles are the male's major internal reproductive structures.

The primary function of the male reproductive system is to produce, transport, and introduce sperm into the female's body. DNA contained within the gametes is responsible for passing on the male's traits to his offspring. The process of producing an adequate quantity of sperm and providing transmission of these gametes out of the male's body involves several important structures, as well as a set of chemical messengers called hormones.

Hormones

Hormones, chemical messengers generated by glands, regulate many body functions. The term "hormone" literally means "to excite" or "to spur on," and hormones often cause things to start happening. As a young person reaches puberty, glands throughout the body begin releasing hormones that cause dramatic changes in body structure and function.

Even though most hormones are found in both males and females, they play slightly different roles in each sex. As a young person matures, luteinizing hormone (LH) is released by the pituitary gland in the brain. The presence of LH in the male body triggers production of the male hormone, testosterone, by interstial cells. Testosterone, along with follicle-stimulating hormone (FSH) from the brain's hypothalamus, stimulates the development of mature sperm from primitive sperm cells.

Once activated, interstitial cells generate relatively stable amounts of testosterone during the entire life span of a male. Testosterone plays several important roles. As a boy reaches manhood, testosterone causes striking changes in his body. The rising levels of testosterone cause a boy's reproductive organs to develop into their adult size. This hormone also initiates development of the male secondary sex characteristics such as deepening of the voice, increased hair growth over the body, enlarged skeletal muscles, thickening of the bones, and increased sex drive.

Without testosterone, male reproductive organs never reach their full size, and adult masculine features such as facial hair do not appear. Since testosterone is produced by interstitial cells, which in turn are located in the testes, without the testes, there can be no testosterone. This is why male castrations, or removal of testes, result in feminizing characteristics.

Spermatogenesis in the Sperm Factory

As an intact (uncastrated) male matures, his body begins to make sperm. Spermatogenesis, the creation of mature sperm, is a process that begins at puberty and continues throughout a man's lifetime. A normal, mature male produces several hundred million sperm each day. Since only one sperm is needed to fertilize an egg, it would appear

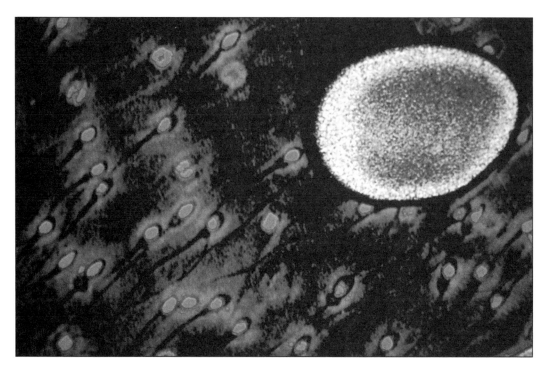

Sperm move through a female's reproductive fluid toward the egg.

that nature has created a generous number of participants in the race to the egg. Most sperm will never make it to their target; a variety of obstacles make the race extremely difficult. Some sperm are not formed correctly, so they never even leave the starting line. Many are killed by the acidic conditions within the female's reproductive system. Of the 100 million that begin the race, only a few hundred ever get close to the egg. The enormous numbers of sperm raise the chances that at least one will make the entire trip successfully.

A sperm develops from a spermatocyte, which descends from a more primitive parent cell. Each spermatocyte undergoes two meiotic divisions to give rise to four new cells. These four new gametic cells, containing twenty-three chromosomes, one-half the genetic material of body cells, are called spermatids.

Spermatids resulting from meiosis are immature sperm. Further changes are required for the sperm to achieve optimal function. It takes a total of sixty-four to seventy-two days for a spermatid to go through the entire development process that creates a mature sperm cell, one that is capable of fertilizing an egg. During this period of maturation, sperm cells become more motile and shed their excess cytoplasm. As sperm mature, they are especially vulnerable to environmental factors. Spermatogenesis can be compromised by factors such as radiation, drugs, alcohol, and tobacco. If sperm are kept too warm, as they can be when tight clothing holds them close to the body, they may not develop properly.

Sperm are some of the smallest cells found within the human body. Each tiny cell has three main parts: a head, a midpiece, and a tail. The head of a mature sperm contains chromosomes plus a supply of enzymes that will eventually help the sperm penetrate an unfertilized egg. The midpiece of the sperm is filled with mitochondria, structures that supply the large amounts of energy the sperm need on their trip toward the egg. The tail of the sperm cell, or flagellum, is designed to move the sperm forward with powerful, whiplike motions.

Housing the Sperm

Spermatogenesis occurs in two organs called testes, the primary male reproductive organs. Testes also play a key role in production of the male hormone, testosterone. All other parts of the male reproductive system are ducts and glands that aid in the delivery of sperm to the female's body.

Each testis in an adult male is about the size of a large olive. These paired organs are suspended outside the body in a protective sac called the scrotum. The scrotum, which normally hangs loosely between the thighs, provides the protection required by any delicate internal organ and also helps to maintain the testes at temperatures necessary for sperm production.

Sperm cannot be manufactured in the hot environment of the male's abdominal cavity. Viable sperm can only be produced within a narrow range of temperatures—not too warm and not too cool. Generally, in their protective sacs, testes remain at their optimal temperature, about three degrees Celsius cooler than body temperature.

However, there are times when the scrotum helps to warm the testes. When the male is exposed to very cold conditions, the scrotum wrinkles and pulls the testes near the body wall. By doing so, the warm male body provides enough heat to promote sperm production.

In an unborn male, testes are initially located inside the body cavity, where they develop during the first and second trimesters of fetal growth. However, in the last trimester, the testes migrate downward through the abdominal cavity until they descend into the sac outside the body. During its descent each testis carries its ducts, blood, lymphatic vessels, and nerve fibers with it.

Frequently, a male baby is born with undescended testes. Usually, the testes will move into their correct positions shortly after birth. If this does not happen by the time a boy

Testosterone is the male hormone responsible for development of such characteristics as a beard, pubic hair, and change of voice.

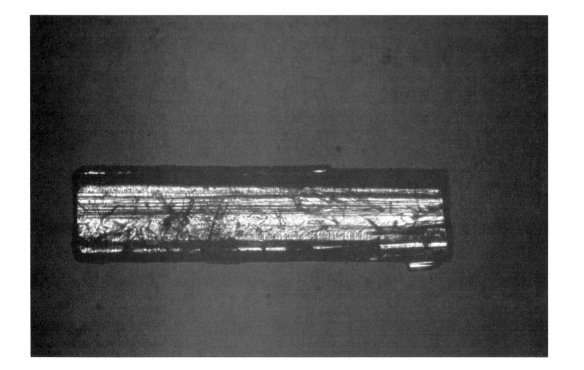

reaches puberty, however, surgery is often performed. Even though a man with undescended testes is able to have sex, he cannot become a father, since healthy sperm production is inhibited inside the warm abdominal cavity.

Inside the scrotum, a fibrous capsule called a tunica albuginea surrounds each testis and partitions it into tiny sections. Each testis has about 250 sections or lobules. Every one of the lobules contains from one to four tightly coiled structures called seminiferous tubules. These tubules serve as the sperm factories of the testes.

Sperm Begin Their Journey

Developing sperm spend a lot of their time in a series of tubes. The sperm-making seminiferous tubules wind around each other to form a tangled network of tubes that are collectively called the rete testis. Once immature sperm are produced in the seminiferous tubules, they are transported through the rete testis to another structure, the epididymis. The epididymis, a very narrow coiled tube that is about twenty feet long, is the beginning of the duct system that eventually carries sperm out of the male's body. The epididymis, which is shaped like a comma, hugs the external surface of the testis. It extends from the top of the testis down its backside.

It is within the epididymis that sperm mature and become motile. The maturation process that started within the testes and culminates in the epididymis yields hundreds of millions of sperm that are completely mature and free swimming. The mature sperm remain in the epididymis until the male becomes sexually stimulated. Once aroused, the walls of the epididymis contract, causing the sperm to travel to their next segment of the journey, the vas deferens.

The muscular vas deferens, also called the ductus deferens, is about eighteen inches long. Lined with cilia that help sweep the sperm along its way, the vas deferens spans the top of the bladder and eventually fuses with another tube to form the ejaculatory duct.

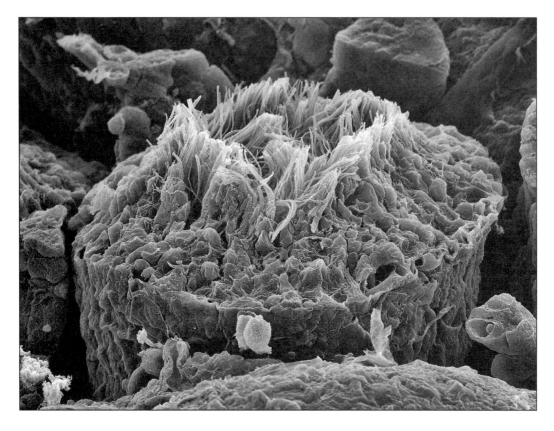

Millions of sperm cell tails swirl out of a sectioned semi-niferous tubule.

Men that do not want to father children sometimes elect to have their vas deferens disconnected. In a simple surgical procedure called a vasectomy, a portion of the vas deferens is cut out and the tubes are tied or sealed so sperm cannot leave the body. This operation does not interfere with testosterone production or sex drive.

Supplies for the Trip

Special secretions are added to the sperm to nourish them and facilitate their passage out of the male's body. These fluids, plus the sperm they support, are collectively called semen. The secretions come from three different sources: the seminal vesicle, the prostate gland, and the bulbourethral gland.

The seminal vesicle, located at the base of the bladder, is the tube that fuses with the vas deferens to form the ejaculatory duct. The seminal vesicle turns out a thick yellow secretion called seminal fluid that forms about 60 percent of semen's volume. Seminal fluid contains sugar, vitamin C, and amino acids that help nourish sperm on their trip.

The prostate gland, about the size of a chestnut, surrounds the urethra just below the bladder. It releases a milky alkaline fluid that is filled with nutrients and enzymes into the urethra. Prostate fluid helps neutralize the acidic conditions of the female's vagina, and without this assistance, few sperm would survive long enough to complete the journey to the egg.

The bulbourethral or Cowper's glands are two pea-sized structures located just below the prostate gland on either side of the urethra. They too add alkaline fluid to neutralize acidic environments, as well as lubricating fluid that facilitates sexual intercourse.

Beyond the entrance of the Cowper's and prostate glands, the ejaculatory duct opens into the urethra. The urethra travels through the length of the penis, and then culminates at an opening at the end of this structure. The penis consists of a skin-covered shaft, which ends in an enlarged tip, the glans. The outer surface of the penis is covered with a loose layer of skin called the foreskin or prepuce. This skin is usually removed from male children by a surgical procedure called circumcision.

Propagating the Species

Males are responsible not only for producing sperm, but also for delivering them. Mature sperm must be deposited in the female's body so they may encounter the female gamete, the egg. The act of placing sperm into the female's body is called coitus, sexual intercourse, or copulation.

The penis must be rigid or erect to be inserted into the vagina, the entrance to the female reproductive system. A nerve reflex that is triggered by a variety of stimuli, including visual, tactile, and mental, controls erection. Inside the penis, the urethra is surrounded by layers of erectile tissue, a type of spongy tissue that contains a lot of blood vessels. During sexual arousal, nerve impulses signal these vessels to dilate, so blood flow to the penis increases. As a result, blood fills the spaces of the spongy tissue. The swelling of these spaces presses against veins in the penis, slowing the return of blood to the heart. Therefore, more blood flows into the penis through the arteries than leaves it through the veins, enabling the penis to enlarge and become rigid.

Once erect, the penis can be placed inside the female's vagina. Arousal results in emission or ejaculation of sperm and semen. During ejaculation, the muscles around the vas deferens contract. These contractions plus contraction of muscles at the base of the penis force semen from the male's body into the female's vaginal canal. Ejaculation is accompanied by increased heart rate, elevated respiration, and higher blood pressure as well as intense sexual excitement. Sperm immediately begin swimming in search of an egg.

An adult male produces about 100 million sperm cells in each milliliter of semen expelled from his body. Most men expel between two and four milliliters of semen with each ejaculation. Of the 300 to 400 million sperm released in the female's vagina, only a few sperm will survive long enough to reach oviducts, where an egg may be present for one or two days each month. It is because of this low survival rate that a high sperm count is required to achieve fertilization. Males that have fewer than 20 million sperm per milliliter of semen are usually considered sterile.

A Common Goal

The goal of any species, including humans, is to perpetuate itself. Living things create a future for their kind by reproducing. Some organisms, like bacteria, reproduce asexually by cloning themselves. More complex living things produce specialized sex cells that give rise to unique offspring.

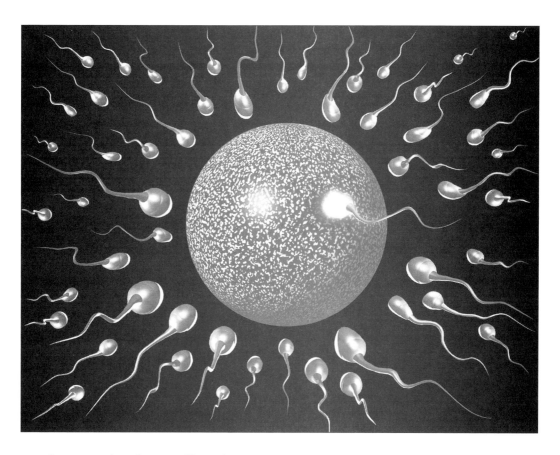

The specialized sex cells in humans are called eggs and sperm. These cells result from a series of cell divisions that reduce the normal chromosome number to half that of the parent cell. With a half-set of genetic material, male and female sex cells unite to form a zygote with a complete set of chromosomes. The organs that produce these cells depend on the other organ systems of the body for nutrition, oxygen, structure, and all other forms of support. Yet the reproductive organs give nothing back to the body except the opportunity to produce offspring.

The male reproductive system is a streamlined, efficient factory that churns out millions of sperm daily. By providing many more cells than are actually needed to fertilize an egg cell, the male reproductive system assures its success. The cells it generates are so specialized that

Artwork shows a sperm cell fertilizing an egg, combining the genetic material necessary to reproduce.

they are nothing more than strands of DNA powered by an engine (the mitochondria) and a tail (the flagellum). When placed inside the female's vagina, these tiny genetic soldiers have one goal in common: finding and fertilizing the female gamete. Even though millions of sperm begin the trip, only a few survive to reach the unfertilized egg. Of those remaining few, only one spermatozoan will be allowed to penetrate and pass its DNA into the egg.

3 Female Reproductive Organs

In ancient times, the processes that brought about the birth of a baby were poorly understood. The Greek teacher Aristotle (384–322 B.C.) believed that a mystical spirit caused the creation of a new life within a woman. He wrote that semen from a male, with the aid of supernatural forces, caused a baby's tiny limbs and organs to be formed from female menstrual blood. Galen (A.D. 129–199), a Greek writer who was also a physician, expanded Aristotle's theory, suggesting that miniature, preformed humans already existed within all females. He believed that contact with a male's semen allowed them to break out of their shells and emerge.

Aristotle, Galen, and other respected teachers believed that pregnancy and childbirth were the "burden" of women, and that menstruation was their "curse." This attitude persisted for centuries. Consequently, little was done to study reproductive processes. It was not until the early 1900s, after development of the microscope showed that pregnancy was caused by the fusion of a male and a female cell, that reproductive organs received much attention from the scientific community.

Today, the functions and structures of the female reproductive system are quite well understood. An explanation of the female's reproductive system is more difficult than that of the male system because of the multiple jobs for which it is responsible. The female system has the duty of producing the female gametes, called eggs, ova, or

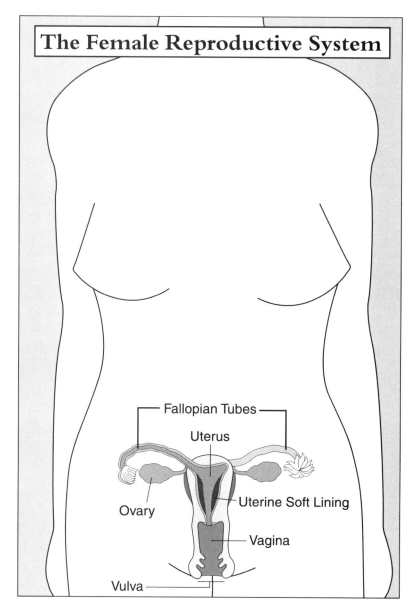

The Female Reproductive System

Fallopian Tubes

Uterus

Ovary

Uterine Soft Lining

Vagina

Vulva

oocytes. This system must also provide a means to move the egg from the place it was made to a location where it can encounter a male reproductive cell if one is available. If the egg successfully unites with the male cell, it eventually becomes a new human being. In the period between the union of sperm and egg and birth, the female reproductive system provides the site in which the developing cell can grow.

The External Reproductive Structures

A woman's reproductive structures can be classified into two categories: external organs and internal organs. As a group, all of the external structures are called the vulva, or the female genitalia. The vulva includes the mons veneris, labia majora, labia minor, clitoris, and vestibule. These organs extend from the pubic area to the rectum, and they surround the openings of the vagina and urethra.

The mons veneris, Latin for "mound of Venus," is an area of fatty tissue below the stomach, lying directly over the pubic bone. Because it contains a lot of nerve endings, the mons is very sensitive. Just below the mons are the labia majora, literally "large lips." These consist of two folds of skin surrounding fatty tissue that are covered on their outer surface with pubic hair. These folds cover and protect the rest of the genitals.

Inside the labia majora are two smaller, hairless folds, the labia minora or "small lips." These fleshy folds are filled with blood vessels and nerve endings. Both sets of labia come together in the front of the body to form a hood of flesh that covers the clitoris. The clitoris, which is about an inch long, is a very sensitive organ made of the same type of erectile tissue found in the male's penis.

Between the labia minora and vagina is the urethra, the opening to the urinary tract, and the vestibule, the area around the opening to the vagina. Two vestibular glands are located on either side of the vaginal opening. These secrete lubricating fluid, similar to the male bulbourethral gland.

Stimulation of the Female Genitalia

All of the tissues of the vulva are sensitive to touch and sexual stimulation. Sexual excitement causes erectile tissue in the clitoris and around the entrance to the vagina to fill with blood. Stimulation triggers nerves that carry impulses to the spinal cord, resulting in dilation of arteries in the genital tissue. These same nerves cause the vestibular glands to produce lubricating fluid near the opening of the vagina to assist with the insertion of the penis. Culmination of stimulation to female genitalia can lead to orgasm, a pleasurable sensation. Orgasm causes muscles of the pelvis to contract, a reflex that helps transport sperm through the female tract.

Internal View of the Female Reproductive Tract

The rest of a female's reproductive organs are located deep inside her body. These include the ovaries, oviducts, uterus, and vagina. The ovaries, oviducts, and uterus create and house eggs, the specialized female reproductive cells.

Ovaries, the organs that produce eggs, are two small oval masses of tissue located in the lower abdominal cavity. Since the ovaries are the source of female gametes, they are comparable to the testes in males. Each ovary measures about one and one-half inches long and an inch wide, about the size and shape of a large almond.

The surface of each ovary is dimpled and puckered. This outer layer contains millions of follicles, cell masses that contain eggs. The inner part of each ovary is largely made up of loose connective tissue that is fed with an ample blood supply.

Making the Eggs

Amazingly, all of the eggs that a female will ever possess develop within her body before she is born. The cells that are destined to become eggs begin forming in each ovary just six weeks after conception. Eggs, also called oocytes (from Latin and Greek words for egg and cell), develop in the granular-looking follicles. Within its own follicle, a blanket of protective follicular cells surrounds each oocyte.

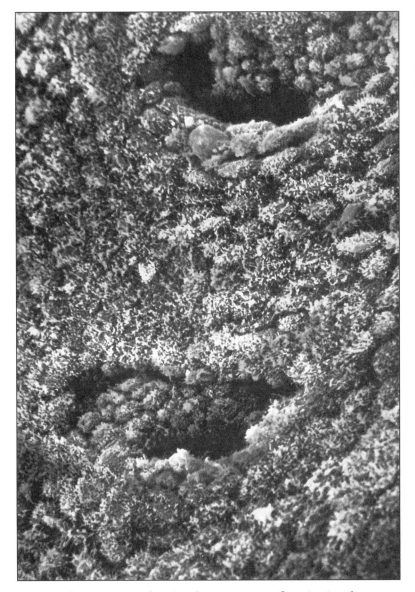

Two depressions in an ovary reveal high meiotic activity during the development of an egg.

Developing eggs begin the process of meiosis, the type of cell division that reduces the number of chromosomes. However, meiosis stops midway through the process, leaving the eggs in a state of suspended animation that lasts for at least ten years.

Once all of the oocytes are formed in a developing female, no new ones ever appear. In fact, the number of oocytes begins gradually declining before a female child is born. At twenty weeks, the ovaries of a female fetus contain 6 to 7 million egg cells. This number has dropped to 1 to 2 million cells at birth, and by the time a female reaches puberty, it has plummeted to only 300,000 eggs. This decline continues throughout the reproductive life of the female. Only about 300 to 500 eggs actually reach maturity and are expelled from the ovary during ovulation.

After a female's birth, the ovaries remain quiet for some time. When a girl reaches puberty, the time in adolescence when changes occur that enable the body to carry out reproduction, some of the oocytes are stimulated. Each month, a few cells finish the process of meiosis that began so many years before. Not all of the oocytes are prompted to mature at one time. Only those that are stimulated develop into full-fledged female gametes, containing only a half-set of chromosomes.

In a cell that is destined to become a sperm, meiosis results in four cells of approximately the same size, but meiotic cell division in females is different. When an oocyte goes through its first meiotic division, it creates two cells of very different sizes. The larger of the two is called the ovum or egg cell. In comparison to other human cells, it is gigantic, the biggest cell found in people of either sex. The second, smaller cell is called a polar body. If the ovum is fertilized, it divides unevenly again. The resulting giant cell is the fertilized egg or zygote. The smaller cell is another polar body. All polar bodies are nonfunctional, and are reabsorbed by the female's body.

Chemical Triggers

The trigger that reactivates some of the oocytes at puberty is a hormone. Hormones are chemical messengers created in one part of the body to regulate functions in another area. As a young girl matures, the pituitary gland in her brain secretes a hormone called FSH, follicle-stimulating hormone. It is the presence of FSH that reactivates meiosis in egg cells. Each month FSH causes division and growth of cells lining several follicles. When follicular cells begin dividing, they fill the sac surrounding the oocyte with fluid that bathes and protects the egg. These sacs enlarge until, at their maximum size of about 0.4 inches, they bulge up from the surface of the ovary like blisters.

As the sacs mature and increase in size, so do the oocytes. By the time the sacs have reached their greatest size, the oocytes have grown into large cells that are surrounded by a protective, jellylike coating called the zona pellucida. Outside the zona pellucida, the layer of follicular cells called the corona radiata serves to nourish and care for the oocyte within. Although twenty or more follicles are stimulated each month, one grows more rapidly than the others. The fastest-growing follicle swells to the point that it finally bursts, releasing its precious contents. The release of the egg from its follicle is called ovulation. The other oocyte-containing follicles disintegrate.

Passageways to the Uterus

When the egg bursts free, it is released into the pelvic cavity. Ejection from the follicle sends the egg floating toward the mouth of a tube that leads to the uterus. The ovaries are connected to the uterus by Fallopian tubes or oviducts about five inches long; there is one oviduct on either side of the pelvic cavity.

At the ovary end of each oviduct, there is a gap between the oviduct itself and the ovary. To compensate for this gap, ensuring that the egg enters the oviduct and does not get lost in the pelvic cavity, the tube's opening is enlarged and shaped like a funnel. Around the edges of the funnel is a cellular fringe of long, thin fibers that sweep the egg toward the oviduct. Despite these aids, some eggs never find their way into the passageway. If an egg accidentally slips into the pelvic cavity, it may be fertilized there, leading to a dangerous condition called an ectopic pregnancy.

The inside of each oviduct is lined with cells that produce mucus. Many of these cells are topped with tiny hairlike

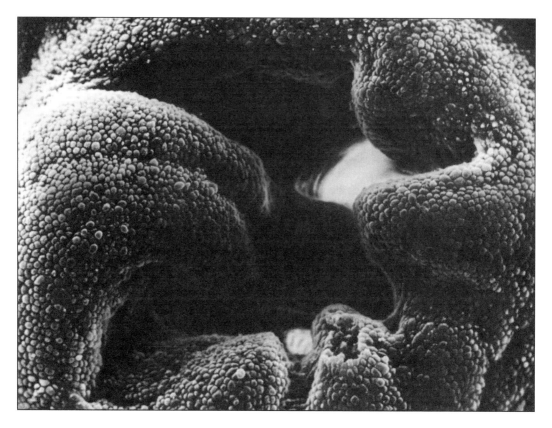

The oviduct carries an egg from the ovary to the uterus.

extensions called cilia. The presence of mucus, along with the gentle sweeping movement of cilia, guides the egg through the oviduct on its way to the uterus.

Once an egg has left the ovary, it can only live for a period of thirty-six to forty-eight hours. After this time, it breaks apart, ending the potential for fertilization. However, if sperm are present right after the egg is released, one spermatozoan may unite with the egg as it begins its trip down the oviduct. The union of the egg and sperm is called fertilization.

To prevent pregnancy, some women "have their tubes tied," the tubes being the Fallopian tubes or oviducts. In a relatively simple operation called a tubal ligation, the oviducts are clipped and their free ends knotted. This surgical intervention eliminates the site where fertilization usually occurs, and prevents eggs from traveling to the uterus.

The Cradle of Development

The uterus is a hollow, muscular organ located below and between the ovaries. To accommodate a developing fetus, the uterus can enlarge as much as twenty times its original size. In the nonpregnant condition it is about the size and shape of a small pear. From the front, the uterus looks like a cow's face; the oviducts resemble horns while the lower end looks like a nose. If a fertilized egg reaches the uterus, in a healthy pregnancy it will remain there until development is complete.

The lower end of the uterus is called the cervix. This tubular area operates like a gate between the uterus and the vagina. The vagina is a muscular tube about four inches in length that joins the uterus to the outside of the body. The vagina has several important jobs. Secretions from the uterus leave the body via the vagina. During sexual intercourse, the vagina receives the penis. At birth, the vagina is the passage through which the baby leaves its mother's body.

Passage to the Outside

At birth, the opening of the vagina is somewhat closed by a thin membrane called the hymen. The hymen is not the same in all females. It may appear as a ring or as a series of separate bands that partially cover the opening. Some females are born with only a partial hymen, whereas others have none at all. The first time a penis enters the vagina, it may tear an intact hymen. However, the presence or absence of this membrane cannot be used as an indicator of virginity, since some females never possess one and those that do may prematurely tear it through insertion of tampons, vigorous physical exercise, or injury.

The Hormone Control Panel

Hormones control all of the changes that prepare the female body for reproduction. The hormone androgen begins circulating through a girl's body as early as six years of age. Over time, blood levels of androgen reach adequate levels

Two teenage girls look at a calculator. Girls begin puberty between the ages of eleven and fourteen.

to cause development of hair under the arms and in the pubic area. About the same time, the hormone estrogen stimulates development of the breasts, and prepares the body for the first menstruation. Puberty, which generally occurs between the ages of eleven and fourteen years, is the time in a girl's life when ovulation and menstruation begin.

The hormones that dictate a female's reproductive life also provide monthly messages and directions. These chemical messengers come from a variety of sources. Some are secreted by the hypothalamus, a gland at the base of the brain; some from the pituitary, a gland within the brain; and others from the ovaries.

Before the age of eight, a female is reproductively immature. Shortly afterward, the hypothalamus stimulates the pituitary to release two hormones, FSH and luteinizing hormone (LH). In females, these hormones control the maturation of sex cells and stimulate the release of the two sex hormones, progesterone and estrogen. These two vital hormones are secreted by the ovaries, by the adrenal cortices (small glands on top of the kidneys), and by the placenta during pregnancy.

Most of the estrogen produced in a nonpregnant female comes from the ovaries. Estrogen causes several changes in a maturing girl's body, including enlargement of the vagina, uterus, oviducts, and ovaries. It also triggers the appearance of secondary sex characteristics, such as development of breasts, and deposition of fat under the skin, especially in breasts, thighs, and hips. In a nonpregnant female, the ovaries also serve as the main source of another hormone, progesterone. Progesterone causes changes in the uterus during the female reproductive cycle. It also plays an important role in maturation of breasts.

Food for a Babe

Although breasts, or mammary glands, are not technically reproductive organs, they do play an important role in human reproduction because they provide food for the newborn.

Each mammary gland is made up of about twenty lobes, which look like clusters of grapes attached to stems that represent ducts. Each lobe contains numerous clusters called alveoli, groups of cells that can secrete milk. A tiny duct that connects to larger ducts drains each alveolus.

Fatty tissue separates the lobes and cushions the alveoli. Eventually, all ducts lead to the nipple. A ring of pigmented skin called the areola surrounds the nipple, a protrusion located near the center of the breast. This area contains lubricating glands that soften the nipples and keep them pliable during nursing.

Even though breasts vary in size from one individual to another, size has nothing to do with their ability to function. Large breasts simply contain more fat than smaller ones. Breasts naturally increase in size during lactation. In non-nursing females, the size and sensitivity of breasts are affected by hormones that regulate the monthly menstrual cycles.

Cyclic Occurrences

The average menstrual cycle lasts about twenty-eight days and has three distinct phases: menses, the follicular (or proliferative) stage, and the luteal (or secretory) stage.

Generally menses last two to seven days. During this time frame, the uterine lining is shed and leaves the body by way of the vagina.

The follicular phase of the menstrual cycle begins after menses when FSH stimulates the ovaries, causing one of them to select about twenty follicles to begin the maturation process. As cells with each follicle develop, they produce estrogen and some progesterone. Production of estrogen early in the follicular phase repairs the lining of the uterus, thickening it from the width of a hair, a slim .02 inches, to the thickness of yarn, about 0.1 inch.

Next secretions from the pituitary cause all follicular cells to increase their output of progesterone. Rising progesterone levels stimulate the pituitary to release LH. As a consequence, LH surges through the body, triggering the rupture of the fastest-growing follicle, which releases its egg cell. This event is called ovulation.

After ovulation, the luteal phase begins. The presence of LH causes changes in the leftover pieces of the ruptured follicle. This ruptured follicle collects fatty mate-

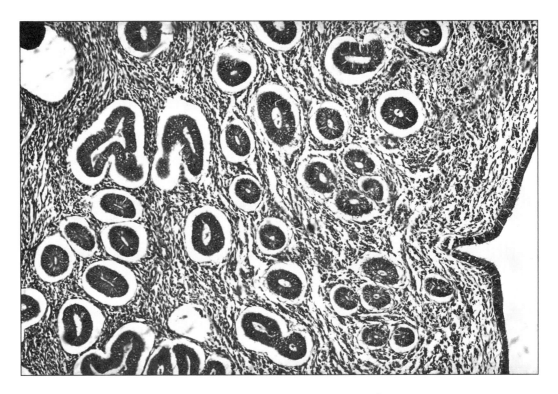

rials, causing it to drastically enlarge and take on a yellow color. The newly formed, temporary glandular structure, called a corpus luteum ("yellow body"), secretes progesterone during the last half of the reproductive cycle. Presence of progesterone further thickens the lining of the uterus to a depth of 0.2 inches, or almost as thick as a pencil. Glands and blood vessels also multiply in the endometrium.

Estrogen repairs the lining of the uterus, increasing the thickness to 0.1 inch.

If the egg is fertilized by a sperm, the female becomes pregnant. If fertilization does not occur after ovulation, however, the corpus luteum begins to disintegrate. Reduction of progesterone in the blood causes the tiny blood vessels in the endometrium to spasm and tighten, cutting off its blood supply. As a result, tissue in the endometrium begins to die and detach from the wall of the uterus. At the end of the luteal phase, a mixture of detached endometrium and blood from the severed capillaries is expelled from the body through the vagina.

The menstrual flow, or the "period," continues anywhere from two to seven days, allowing discharge of the unneeded uterine lining.

The Ovaries Go into Retirement

When a female is about the age of forty, the events of the reproductive cycle become less predictable. At that time, estrogen levels begin to drop, and the ovaries are not stimulated to carry out their normal routine. As a result, the monthly menstrual cycle occurs less frequently. Eventually, ovulation and the menstrual cycle stop completely, and a woman's reproductive life is over. This change, called menopause or climacteric, usually occurs between ages forty-five and fifty-five.

Menopause is a gradual change that takes several years. During this time, levels of progesterone and estrogen drop. The changes that accompany dropping hormone levels can cause discomfort and irritability for some women. One unpleasant symptom menopausal women may experience is hot flashes, feelings of sudden warmth that are often accompanied by sweating and reddening of skin. Low levels of estrogen can also reduce the amount of lubricating fluids that are produced in the vagina and lessen the elastic nature of vaginal tissues.

The Story of Hormones and Cells

For centuries the mechanics of reproduction were poorly understood. Most physicians considered reproduction to be a mystical process that

An older woman plays tennis. Later in life, women undergo menopause, a process that ends their reproductive life.

did not involve the participation of men in any way. Only in the last five hundred years have educated people considered the processes involved in reproduction as science.

Like all living things, humans reproduce to propagate their species. However, human females have very different reproductive cycles than females of other animal species. Whereas most animals produce egg cells only once or twice a year, a sexually mature human female generates one every month for about thirty years.

To make reproduction possible, a woman's body undergoes drastic changes on a regular basis. Relentlessly, month after month, it produces an egg and prepares for pregnancy. A variety of hormones play important roles in controlling and coordinating these changes. Hormones stimulate the ovaries to mature and release an egg cell, and also prepare the uterus to receive a fertilized egg. Hormones also maintain the woman's body in a condition that will nourish the developing offspring. Once a baby is born, the entire series of events begins again.

4 | Fertilization and Development

Babies are born every day all around the world. These familiar yet miraculous events mark the culmination of a series of very complex processes in the human body. A baby begins life when a sperm and egg join to form one fertilized cell. Over the next nine months this one cell develops into an organism composed of trillions of specialized cells.

The first step in the formation of a baby is the union of sperm and egg. For this process, called fertilization, to occur, a sperm must find and penetrate an egg. Females do not produce and release eggs every day. A female ovulates, or releases an egg from one ovary, about the midpoint of her menstrual cycle. The egg bursts from its ruptured follicle and floats into an oviduct, where it remains viable (capable of being fertilized) for only thirty-six to forty-eight hours. If a sperm does not penetrate the egg during that time, it degenerates and is swept from the body during the next menstrual cycle.

Predictions about the times at which an egg is or is not available for fertilization are only generalizations. Since no two women have exactly the same reproductive cycles, it is very difficult to know whether or not coitus at any given time will result in pregnancy.

The Race Is On

Pregnancy is more probable if sexual intercourse occurs near the time of ovulation. With each male ejaculation, hundreds of millions of sperm are released into the woman's vagina where they begin a marathon race against time and each other. Most sperm can only survive from twelve to forty-eight hours, even though a few "super sperm" are still alive after seventy-two hours. These tiny, but energetic, cells must make the arduous trip through the uterus and Fallopian tubes to the egg. It is not impossible for conception to occur outside of these constraints. Pregnancy is most likely, however, if sexual intercourse occurs no more than seventy-two hours before ovulation and no later than twenty-four hours after.

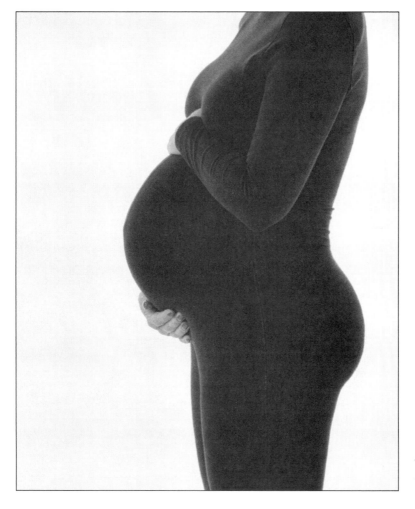

A woman far into pregnancy holds her stomach.

Although hundreds of millions of sperm may begin the trip toward the egg, only a few hundred ever complete the journey. Sperm encounter tremendous obstacles on their trip. Their road map takes them from the vagina, through the cervix, and into the uterus. From there, they travel up the Fallopian tubes.

Sperm push on the uterus wall, attempting to reach the egg.

The first hurdle sperm must overcome is mucus secreted by the cervix. The mucus is thin and most hospitable to the sperm around ovulation. At other times, the mucus is much thicker, making sperm navigation more difficult. The sperm that successfully travel through the mucus are confronted with two passageways at the other end of the uterus. Some sperm dash up one Fallopian tube while others choose the other tube. Since there is only one waiting egg, about one-half of the surviving sperm take a wrong turn and enter the empty Fallopian tube.

Powered by mitochondria, strong flagella whip the sperm that took the correct route up the Fallopian tube to the goal. Since the egg does not travel even halfway down the Fallopian tube during its viable period, the sperm have several inches of oviduct to navigate. The amount of time it takes sperm to travel from the vagina to the egg varies. Some can make the journey in as little as fifteen to twenty minutes, while others may require hours.

The Prize Does Not Always Go to the Swiftest

Eventually several hundred sperm make it to the final stretch of their journey and find the waiting egg. In a frenzied mass, the sperm surround the egg, pushing against it with their heads. In doing so, they release enzymes, which begin dissolving the egg's protective outer membrane. Eventually, the membrane is weakened enough to allow one sperm to penetrate. The successful spermatozoan is usually not the swiftest: After the repeated bombardment by hundreds of early arrivers, a latecomer may be the one to break into the egg.

Once the egg is penetrated by one spermatozoan, its cell membrane undergoes some changes. It sends out an electrical signal that causes cortical granules, small sacs beneath the membrane, to release their contents into the open space surrounding the egg. As a result, the egg begins to swell and other sperm are pushed away. All sperm except the one that made its way inside the egg die within the next twenty-four to forty-eight hours.

The changes in the egg caused by sperm penetration trigger the last round of meiotic cell division in the egg's nucleus. At last, a nucleus is formed in the egg that contains the female's twenty-three chromosomes. Within the next twelve hours, the nuclear material from the head of the sperm, which contains the male parent's twenty-three chromosomes, migrates to meet the nucleus of the egg. The two nuclei fuse to form one cell that contains a full complement of forty-six chromosomes. This cell, which contains the genetic material of both parents, is referred to as a zygote, the first cell of a new individual.

The Journey of the Zygote

Shortly after formation, the zygote starts undergoing mitotic cell divisions, a process that will continue until the zygote develops into a full-term baby, ready to be delivered. As it begins dividing, the zygote is gently swept down the Fallopian tube toward the uterus. Inside the Fallopian tube, the zygote is helped along its way by slick mucous membranes, contractions of smooth muscles, and the waving action of cilia.

About eighteen hours after fertilization, the zygote splits into two cells, each containing the same chromosome number and identical DNA. Eventually the two cells split to form four. The four cells double to form eight, and the eight double into sixteen. Each doubling of cells in this early stage of embryonic development is referred to as a cleavage. A series of cleavages creates a large number of cells that will eventually give rise to the human embryo.

Home at Last

Within three days after fertilization, the dividing zygote has changed into a solid ball of cells called a morula. The morula, resembling a microscopic raspberry, floats freely within the uterus for two or three additional days as the uterus prepares for its implantation. During this time the morula continues to undergo mitotic divisions, and assumes the form of a hollow ball of cells called a blastocyst. The blastocyst, composed of from 128 to 256 cells, consists of an outer layer of cells surrounding an inner cell mass. The outer layer of cells, referred to as a trophoblast, will eventually become part of the placenta. The trophoblast of the blastocyst burrows or implants itself in the endometrium, the soft, velvety lining of the uterus, seven to nine days after fertilization. It is the inner cell mass of the blastocyst that will develop into an embryo, and eventually form a fetus.

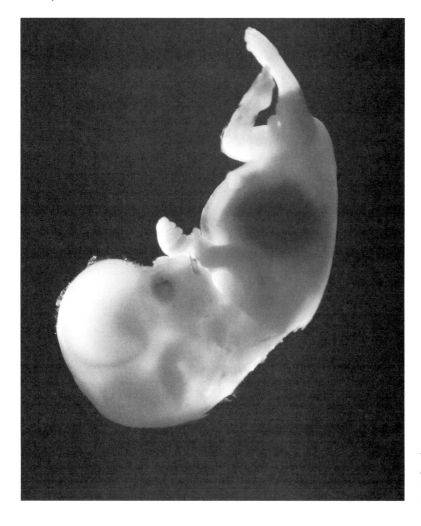

A human fetus forms from the inner cell mass of the blastocyst.

The blastocyst normally begins implantation within five to eight days after fertilization and completes it by the tenth day. By fourteen days after fertilization, the uterine lining envelops the implanted blastocyst, which is no bigger than a grain of sand. This begins a period of development called the embryonic stage.

The Embryo Develops

Early in the embryonic stage, the inner cells of the blastocyst, those that will develop into a baby, form a distinct outer and inner cell layer. A middle cell layer, the mesoderm, soon separates the outer, also called the ectoderm, and the inner layer or endoderm. These three so-called germ layers are responsible for forming all body organs of the unborn child. The ectoderm layer will develop into the nervous system, the sensory organs, the hair, nails, skin, and linings of the mouth and anus. The mesoderm layer will give rise to the muscles, bone tissue and marrow, blood, blood vessels, lymphatic vessels, connective tissue, internal reproductive organs, and kidneys. The cells of the endoderm will form the digestive tract, respiratory system, urinary bladder, and urethra.

As the germ layers are emerging from the blastocyst, another important reproductive structure, the placenta, is also taking shape from the trophoblast. The placenta is not part of the baby-to-be. Instead, it is formed from both cells of the trophoblast and cells of the mother's uterine wall. The placenta develops elaborate, fingerlike projections called chorionic villi that dig into the uterine lining. These fingers securely anchor the placenta to the uterus. They grow alongside the mother's own blood vessels, but they do not penetrate them.

While the placenta is forming, another membrane, the fluid-filled amnion, grows around the embryo. The amnion is connected to a membrane that lines the chorionic villi by a small stalk of tissue. This stalk eventually becomes the umbilical cord. The umbilical cord is made of three blood vessels: two arteries and one vein. It is through these vessels that blood passes from mother to unborn child.

By week three the placenta begins functioning. That is, oxygen and nutrients pass through it from the maternal blood into the embryonic blood, and wastes and carbon dioxide diffuse from the blood of the embryo into the maternal blood. At week three, the embryo is about the size of a raisin. Its heart, eyes, and foundations of the brain and spinal cord have begun to develop.

Week four marks the beating of the heart of the embryo. By the end of this week, muscles are developing and tiny limb buds that will later become arms and legs have appeared. By week five the embryo has grown to one-fourth of an inch, an incredible ten thousand times its original size. During this week facial features, such as the mouth and tongue, form. The major muscle systems have developed, and the embryo has established its own distinctive blood type as its liver is now making blood cells.

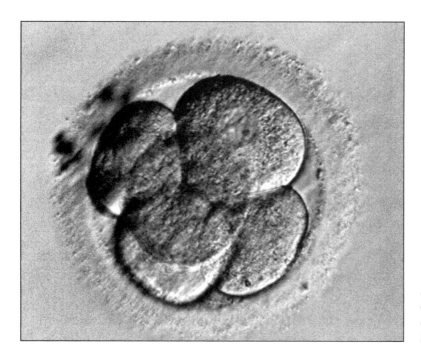

Early in the embryonic stage distinct cell layers are visible in this egg.

By week eight of a normal pregnancy, all the organ systems and skeletal systems have been established, the heart is beating sturdily, and the embryo looks distinctly human. At this point the embryo is about 1.2 inches in length from the crown of the head to rump and weighs about 0.03 ounces.

The Embryo Becomes a Fetus

By week nine, the embryonic period comes to a close and the embryo is officially called a fetus. The head is almost as large as the body, and all the limbs are present. With the foundation in place, the fetus begins a period of growth, organ specialization, and adjustment of body proportions. Weeks nine through twelve mark the nearly complete development of the heart, the appearance of baby teeth in the gums, and full formation of the brain. During this time frame, the fetus sucks its thumb, makes facial expressions, urinates, kicks its feet, moves its hands, and "breathes" amniotic fluid.

During weeks thirteen through sixteen, the fetus, now about five and one-half inches long, develops a very human-looking face, with eyebrows and eyelashes. The body of the fetus is covered by a fine layer of hair. The vocal cords are fully developed, but the fetus is unable to cry, since no air is available. Sex organs are apparent at this stage. The fetus can grasp one hand with another and turn somersaults.

During weeks seventeen through twenty, the fetus can hear and recognizes familiar voices. It is also responsive to stimuli such as light. If bright light is focused on the abdomen of a pregnant woman during this time, the baby moves its hand to shield the eyes from light. Loud music has also been shown to stimulate the fetus to cover the ears. Also during this time frame the unborn child assumes the fetal position, since it is now too large for a full-length stretch. The limbs of the fetus, which move about regularly, are nearing their final proportions.

Weeks twenty-one through thirty are marked by a substantial weight increase as the fetus achieves a crown to rump length of about eleven inches. The fetal body is slender and well proportioned, and the skin is red and wrinkled. The eyes of the fetus are open and the fingernails are now present. During this time the child continues to inhale amniotic fluid into its developing lungs. Research shows that the fetus sleeps about 90 percent of the day and often experiences rapid eye movement (REM) sleep, a stage that is indicative of dreaming. A child born at this stage is said to be premature, but there is a good chance of survival.

During weeks thirty to forty, the fetal crown to rump length reaches fourteen to sixteen inches, and its weight ranges somewhere between six to ten pounds. The skin is pink and fat has been deposited in the tissue under the skin. The birth of the baby is now eminent.

The Mother's Perspective of Pregnancy

Pregnancy is defined as the time from conception to birth. It typically continues for about forty weeks or nine calendar months. Pregnancy is usually measured from the beginning of the mother's last menstrual cycle and terminates with the birth of the child. By 270 days after fertilization, a fetus is considered to be full term and is ready to be born.

A woman goes through some distinct physiological changes in her body during pregnancy. For the first few months many women experience nausea. This "morning sickness" continues until the woman's body has adjusted to the elevated levels of estrogen that come with pregnancy.

By sixteen weeks into pregnancy the uterus has enlarged to occupy most of the pelvic area. As pregnancy ensues, the uterus pushes higher and higher into the abdominal region. This may cause a pregnant woman to experience heartburn as the esophagus is displaced by the crowded conditions within her body cavity.

During pregnancy a woman produces more urine than normal because her kidneys have assumed the additional

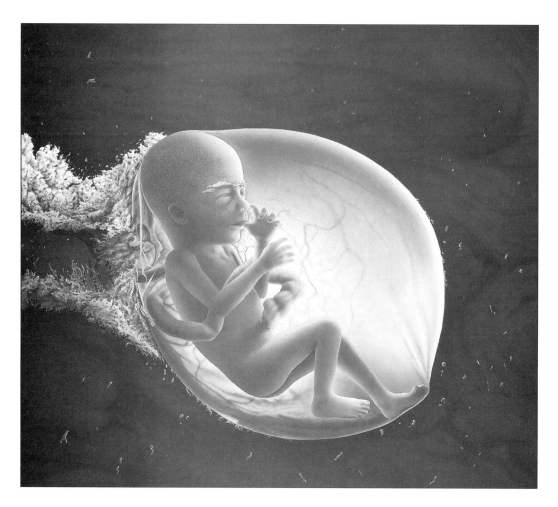

At thirty-eight weeks, a human fetus is nearly fully developed.

burden of disposing of fetal metabolic waste. The growing uterus also presses on the bladder, causing more frequent urges to urinate.

The circulatory system undergoes some dramatic changes. During pregnancy, blood volume rises by 40 percent. Blood pressure, pulse, and cardiac output increase to help propel blood more easily around the body. The uterus sometimes presses on pelvic blood vessels, impairing venous blood flow from the lower limbs.

Some women experience significant weight gain during pregnancy, and many believe that they must now "eat for two." Actually the developing fetus needs only an additional three hundred calories to sustain proper development. The mothers-to-be are advised to eat high-quality, nutritional foods rather than larger quantities of food. Women during pregnancy should avoid ingestion of any harmful substances, since these can cross the placental barrier into the blood of the fetus. Alcohol, nicotine, and drugs, in particular, have been shown to cause damage to fetuses. Viral infections suffered by the mother, such as German measles, can also impair the health of a developing fetus.

Nearing the End of Pregnancy

As the end of pregnancy nears, the woman begins to experience some other changes in her body. The increasing bulk of the abdomen can change the woman's center of gravity, causing her spine to develop an awkward curvature. This may cause backaches during the final months of pregnancy.

As the woman's reproductive system prepares her to give birth, the placenta produces a hormone called relaxin that causes the pelvic ligaments and pubic symphysis to relax and widen, giving them needed flexibility during the birth process.

"A Labor of Love"

Childbirth ordinarily occurs within fifteen days of the calculated due date. The events that lead up to childbirth are called labor. The precise mechanisms that trigger labor are unclear, but a number of factors contribute to the beginning of this process. The hormone oxytocin plays several important roles.

By the last few weeks of pregnancy, estrogen levels in the mother's blood are especially high. Elevated estrogen causes the muscles of the uterus to be very sensitive to oxytocin from the pituitary gland. Initially, the response of muscles to oxytocin sets off weak and irregular uterine spasms called Braxton-Hicks contractions, which pregnant women sometimes misinterpret as "labor." Often referred to as "false labor" pains, they can cause early trips to the hospital.

As childbirth gets even closer, a change in chemical signals converts "false labor" pains into true labor. Oxytocin produced by the fetus itself causes the placenta to release hormonelike chemicals called prostaglandins, which induce the uterus to contract, forcing the baby deeper into the pelvis. Emotional and physical stresses

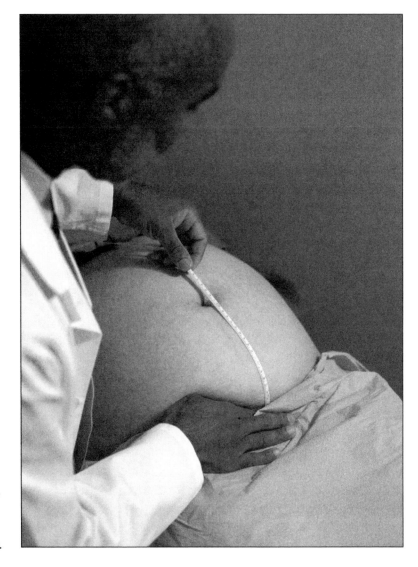

Near the end of a woman's pregnancy her doctor measures the increased size of her abdomen.

now experienced by the pregnant woman cause her hypothalamus to signal additional oxytocin release from her pituitary gland. The combined effects of oxytocin and prostaglandins bring on the contractions of true labor. As the contractions strengthen, even more oxytocin is produced, triggering even stronger contractions. It is important that a woman in labor avoid medications like aspirin or ibuprofen, which can interfere with the actions of oxytocin and prostaglandins.

The Time Has Come

The actual process of labor is divided into three distinct stages: dilation, expulsion, and placental. Dilation, the longest stage, can take from six to twelve hours or even longer. This stage begins when contractions first appear, and it ends when the cervix is fully dilated. Dilation refers to the widening of the cervix to eventually allow passage of the baby from the uterus, through the vagina, to the outside. As the dilation phase begins, the cervix slowly opens a centimeter at a time. Contractions get progressively stronger and occur more often. Each contraction pushes the baby's head against the cervix, which responds by softening and dilating further. In this stage, the amniotic sac around the baby ruptures, an event often described with the phrase "the water has broken."

Stage two of labor, the expulsion, can take up to two hours, although it typically lasts fifty minutes for women giving birth to their first child or twenty minutes for subsequent children. This stage begins when the cervix is fully dilated to ten centimeters or more, and it continues until the infant has passed through the cervix and vagina to outside the mother's body.

A headfirst delivery of the baby is the easiest and safest type. This allows doctors to suction mucus from the baby's mouth, which clears the airways and helps the baby to breathe. If the expulsion stage takes too long, the baby may suffer from oxygen deprivation and be in danger of brain damage.

The last stage of delivery is the placental stage. This usually takes place within fifteen minutes of the complete emergence of the baby, and the clamping and cutting of the umbilical cord. Then strong contractions of the uterus force the placenta out of the woman's body. The placenta and the attached fetal membrane are called the "afterbirth." A small tug on the maternal end of the umbilical cord can help to expel the afterbirth. It is very important for a physician to get all afterbirth fragments out of the woman's body to prevent postdelivery bleeding and other complications.

If a woman is unable to begin or complete a vaginal delivery, the baby is surgically removed from her body. In this procedure, called a cesarean section or C-section, the obstetrician makes an incision through the woman's abdomen and uterus and lifts the baby out.

A doctor holds a newborn baby covered in afterbirth.

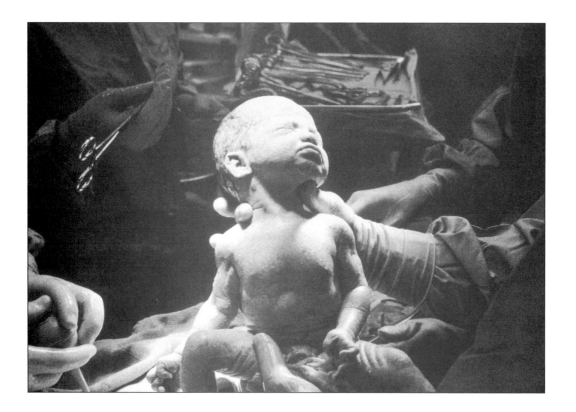

Milk Production

An important event that accompanies the final stages of pregnancy is lactation, or the secretion of milk. Near the end of pregnancy, the pituitary gland stimulates the mammary glands to produce the hormone prolactin, which in turn causes the mammary glands to manufacture milk. For those mothers who choose to nurse, milk serves as nourishment for the newborn.

In the latter stages of pregnancy, the mammary glands also supply colostrum, a material that is abundantly supplied with immune cells. Colostrum is the first liquid secreted from the mother's breasts to her newborn. Colostrum is produced and fed to the newborn for the first two days after birth. The immune bodies in the colostrum help the baby fight off infections at this tender age. On day three, the woman's mammary glands convert to milk secretion. Milk production normally continues for six to nine months, but can last much longer.

The Baby Arrives

The miracle of childbirth has been repeated countless times since humans first appeared on Earth. Yet this incredible occurrence depends on a series of complex and amazing steps. The development of a healthy child hinges on the successful completion of each of these events that take place within the woman's body.

A baby has its beginnings in the gametes, egg and sperm, produced by adult males and females. When a sperm fertilizes an egg, a zygote is formed. This cell contains all the information it needs to make a unique individual. Once formed, a zygote settles into the uterine lining that is designed to support and feed it. There it grows over a period of nine months into a new human being that is capable of living outside of its mother's body.

Childbirth is the process that removes the mature baby from its mother's womb. When the time is right, the mother's body goes through a series of changes that are controlled by hormones. Strong contractions of the uterus push the baby against the closed cervix. Eventually, the cervix begins to relax and widen, providing a way for the baby to escape the only home it has known. After the baby is born, the placenta and other membranes that supported it while inside the uterus are expelled.

Childbirth is a natural process that women have experienced since the beginning of time. In the past, strong, healthy women worked in the fields and homes until time for birth. Then, they paused just long enough to deliver their new child before continuing with their work. Today, women are generally better nourished but not as strong and fit as their ancestors. Therefore, they produce larger babies and often choose to do so in a hospital setting. This makes it possible for the mother or baby to receive medical help if it is needed.

5 Diseases of the Reproductive System

Before the time of the Roman Empire, most doctors and other scientists did not consider a woman's reproductive system to be a topic that warranted study. There were no obstetricians, or doctors who specialize in delivering babies, and no gynecologists, physicians who focus on the female reproductive system. Women depended on family members or midwives to help them during childbirth.

The first great gynecologist and obstetrician, Soranus of Ephesus (A.D. 98–138), lived and practiced medicine in Rome. He helped women with many of the same problems they face today. He gave advice on reliable techniques of contraception, but was opposed to the practice of abortion. Soranus developed new methods to help women through difficult deliveries, and he also taught families how to take care of the newborn. His writings on diseases of women's reproductive organs were considered to be the authoritative source of information for the next fifteen centuries.

Today, both the male and female reproductive systems are subjects of much scientific study. Physicians are aware of many diseases that either originate in or target the reproductive organs. Some diseases are exclusive to gender, while others can strike both sexes.

Problems with Pregnancy

Although most pregnancies result in normal childbirth and delivery, a number of women experience complications

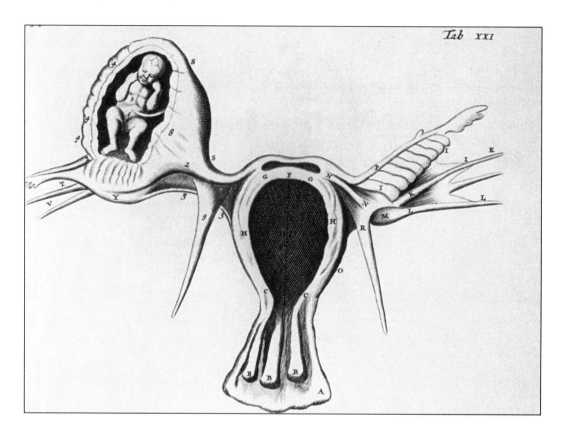

Tab XXI

An illustration of an ectopic pregnancy shows the fetus growing in the fallopian tube.

that result in loss of the fetus. Three of these serious conditions are miscarriage, stillbirth, and ectopic pregnancy.

The term miscarriage is used to describe the loss of a fetus before the twentieth week of pregnancy, whereas stillbirth refers to loss of a fetus after twenty weeks. In most cases, such loss is due to abnormalities in the developing fetus. Bleeding and cramping may be early warning signs of impending fetal loss. A woman who experiences such symptoms generally reports them immediately to her obstetrician, who advises the patient to rest in bed as much as possible.

Normally after an egg is fertilized in the Fallopian tubes, it travels to the uterus where it grows and develops. In an ectopic pregnancy, the fetus develops outside of the uterus, most commonly in the Fallopian tubes. Symptoms are bleeding and cramping, usually after a missed or late menstrual period. Bleeding occurs because the fetus is not implanted in the uterus. Therefore, the uterine lining is shed with the next menstrual flow.

If the fetus dies while very small, the Fallopian tube is not usually damaged. However, if it grows to a large size, the tube will stretch and eventually tear open. Pain and bleeding can result. Bleeding is often gradual, causing an accumulation of blood in the abdominal cavity. However, if the rupture is large, bleeding can be massive, causing the woman to go into shock, a serious medical condition. Without the support of the uterus, a fertilized egg has no chance of developing into a viable fetus. Usually ectopic pregnancies are terminated by the surgical removal of the fetus.

Sexually Transmitted Diseases Caused by Viruses

The reproductive system is the prime target of sexually transmitted diseases (STDs), also called venereal diseases (VDs), but not all diseases of the reproductive tract are the result of sexual encounters. However, STDs are often responsible for the development of very serious conditions such as inflammation of reproductive organs, cancer, and even infertility. Each year in the United States there are 15 million new cases of STDs. Individuals between the ages of fifteen and twenty-four years account for 3 million of these cases.

Sexual transmission of diseases occurs when microorganisms enter the body through moist mucous membranes such as those in the urethra, vagina, mouth, or rectum as a result of sexual intercourse or other sexual contact. Bacteria, fungi, protists, and viruses are examples of microscopic organisms that can cause STDs. Most STDs respond to medication. However, if left untreated, some STDs can cause serious health problems or result in death.

The most common sexually transmitted disease in the United States is genital warts. This condition, caused by the human papilloma virus (HPV), is found in about one-third of all women in the United States. Men and women infected with this disease develop white or pink cauliflowerlike growths in the genital area within a month or two after exposure to the virus. These growths usually occur in clusters that vary in size from very small spots to extremely large, multilobed structures. Although most genital warts are painless, their presence can be the forerunner to more serious diseases such as cancer of the cervix. In fact, nine out of ten women that develop cervical cancer had been exposed to HPV and have experienced a previous bout with genital warts. Men who develop genital warts run the risk of cancer of the penis.

Doctors are usually able to diagnose genital warts by visual exam. If there is a question about a wartlike growth, it can be removed and examined under the microscope to make a positive identification. Although unsightly or annoying warts can be eliminated with liquid nitrogen or laser surgery, the condition tends to recur. Thus, physicians tend not to recommend such removal, but they usually advise women with a history of genital warts to schedule regular pelvic exams to monitor for cervical cancer.

The herpes simplex 2 virus causes genital herpes, another STD. Seven days after contact with the virus, small, painful blisters may appear around the genital area. The pain varies among individuals from mild to excruciating. Two to three weeks after formation, the blisters usually burst and release their liquid contents, which may harden into a crust. When the blisters heal, they leave behind small scars. Herpes simplex 2 virus is usually transmitted from one person to the next by contact with the blister or its liquid contents. Occasionally a person can acquire the virus by touching articles of clothing used by the infected person.

Most of the time doctors are able to diagnose genital herpes by visual exam. When diagnosis is difficult, doctors can scrape some cells from the blister. These cells can be examined under the microscope or used to grow a cul-

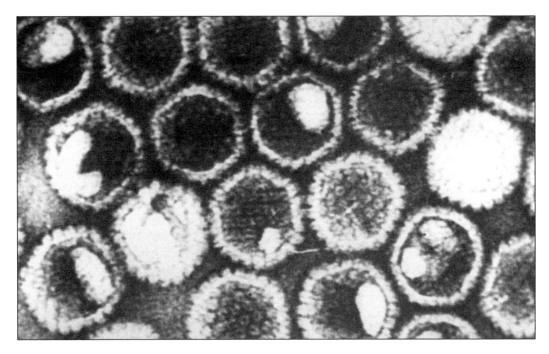

ture of the infectious organism. Laboratory technicians can make a conclusive identification of the causative organism by examination of the culture. As with other viral diseases, there is no cure for genital herpes. However, there are medications that help reduce the pain. The antiviral drug acyclovir slows the viral growth and lessens discomfort. Like genital warts, genital herpes can clear up and then recur. Even when the symptoms are absent, the virus lies dormant in the nerve cells of the body. Factors such as stress can cause the virus to become active again. More than 60 percent of people that get herpes simplex 2 suffer from repeated episodes of this disease.

Herpes simplex virus can be seen under an electron microscope.

Debilitating Bacterial STDs

Bacteria can cause painful, distressing, sometimes fatal sexually transmitted diseases. Syphilis begins with an infection of the bacterium *Treponema pallidum*. This microorganism usually enters the body during sexual relations, but it can also get in through breaks in the skin during nonsexual activities. Once inside the body, the bacteria begin reproducing and travel in the blood and lymph systems to different parts of the body.

Syphilis has three distinct phases. A painless ulcer, called a chancre, is characteristic of the first stage. The chancre forms at the place on the body where the bacteria first entered. It usually goes away within two months. The second stage manifests with muscle weakness, fever, and a skin rash over much of the body. These symptoms may last up to two months before they disappear. For a reason unknown to science, about 70 percent of people with syphilis do not progress to stage three. The 30 percent of people who advance to stage three experience severe symptoms such as swollen joints, inflammation of the lining of the heart, brain, or eyes, paralysis, insanity, and possibly death.

Doctors diagnose syphilis using blood tests that can detect antibodies to the causative bacteria. Since the 1940s the antibiotic penicillin has been used to effectively treat syphilis in stages one and two. However, once a person has progressed to stage three, there is no medication that can reverse the symptoms.

Unfortunately, pregnant women can pass this disease to their babies during delivery. Twenty-five percent of all babies born to mothers with syphilis are stillborn.

Gonorrhea, caused by the bacterium *Neisseria gonorrhoeae,* is another dangerous STD. *N. gonorrhoeae* invades the human body through the lining of the cervix, urethra, rectum, or pharynx following sexual relations. Infected individuals often experience a thick, gray discharge from the cervix or penis that may be accompanied by painful urination. Many women later develop pelvic inflammatory disease, a condition that can scar the Fallopian tubes, perhaps resulting in infertility. If left untreated in males or females, gonorrhea can spread through the bloodstream and bring about inflammation of the heart or brain. If the bacteria find their way to the eyes, the person may become blind.

To form a diagnosis, doctors take samples of the genital discharge and send it to a medical laboratory. In the lab, the samples are cultivated in nutrient-filled flasks. This technique helps analysts to identify the bacteria causing the infection. Once the diagnosis is confirmed, patients are usually treated with oral antibiotics; those with heart and brain complications may need to rest in the hospital while they receive antibiotics intravenously. Since pregnant women with gonorrhea can pass the bacteria to the fetus during delivery, newborns of infected mothers are at risk of developing serious eye infections. For this reason antibiotic ointments are routinely put in the eyes of newborns born to infected mothers.

Annoying Bacterial STDs

Infections of the urethra and cervix can also be caused by a small bacterium called *Chlamydia trachomatis.* This bacterium accounts for about 50 percent of all urethral infections in men that are not caused by gonorrhea. The bacterium is also responsible for many of the pus-forming

Severely inflamed tissue around the eye can result if gonorrhea is untreated and spreads through the body.

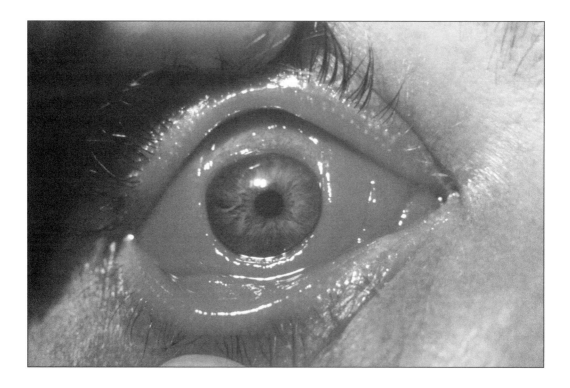

infections of a woman's cervix that are not caused by gonorrhea.

The symptoms produced by *chlamydia* develop between one week and one month after exposure. Infection is usually a result of a sexual encounter with an infected person. Men may be unaware of infection, but those who do experience symptoms feel a mild burning sensation upon urination. Some discharge from the penis is not uncommon. Not all women have symptoms, but those who do notice urinary discomfort, mild abdominal pain, and a yellow vaginal discharge.

Doctors diagnose chlamydial infection by examining the discharge from the penis or the vagina under the microscope. In two-thirds of the people infected with this bacterium, symptoms spontaneously disappear in about a month and serious medical issues do not arise. However, in some people, infection leads to complications. If untreated, a chlamydial infection in women can move into the Fallopian tubes and scar them, a complication that can cause sterility. Men may develop inflammation of the epididymis, which causes painful swelling on one or both sides of the scrotum. Treatment for both men and women includes the use of antibiotics such as tetracycline.

STDs Caused by Protists

The most common STD found in sexually active young women, trichomoniasis, is caused by *Trichomonas vaginalis*. "Trich" is a protist, a microorganism considerably larger and more complex than a virus or a bacterium.

Each year there are about 5 million new cases of trichomoniasis in men and women in the United States. Although usually spread by sexual contact, it can be transmitted by contaminated objects such as washcloths and towels. Even though some women have no symptoms, others experience a foamy green discharge, pain during urination, and genital discomfort. In men, discomfort may result from swelling of the urethra, lesions on the penis, and urinary pain.

As with gonorrhea, trichomoniasis can be cultured from genital discharge. Once the medical laboratory positively identifies the organism, the disease can be treated with metronidazole, a medication that kills the disease-causing protist.

Disease Prevention

Sexually transmitted diseases are easier to prevent than treat. Complete abstinence from sexual relations is the only way to guarantee safety from STDs. Monogamous relationships, in which sexual relations are limited to one person, can reduce risk. Safe sex practices, such as wearing condoms during intercourse, also help to guard against STDs.

In some cases, sexually transmitted diseases can be the precursors to chronic conditions. Many women with a history of STDs develop pelvic inflammatory disease (PID). Statistics show that one out of every ten sexually

Condoms help prevent STDs but they are not 100 percent effective.

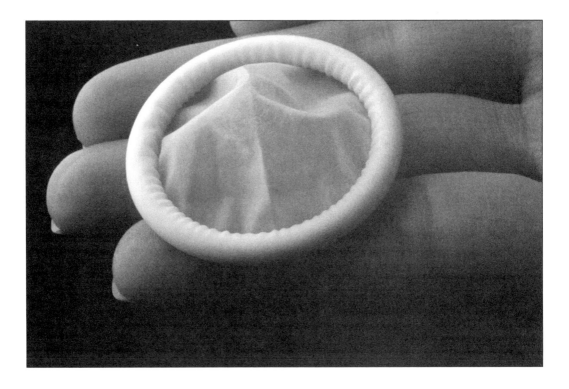

active teenagers suffers from PID. In the United States, untreated cases of bacterial STDs account for 90 percent of all cases of PID. The other 10 percent usually result from infection introduced during an invasive surgical procedure. Most often, however, infecting microorganisms enter the cervix and spread upward to the uterus, Fallopian tubes, and sometimes the ovaries. Some common symptoms of PID include high fever, back and abdominal pain, nausea, vomiting, rapid pulse, chills, and vaginal discharge. Symptoms usually manifest themselves immediately after the menstrual cycle.

Prompt treatment is required for women with PID. As the pathological microbe spreads through the reproductive tract, infection of the Fallopian tubes may occur. Once infected, Fallopian tubes can become blocked by an accumulation of fluid. Left untreated, this condition can lead to sterility due to scar formation in the Fallopian tubes. Each year in the United States, one hundred thousand women become infertile because of PID. Another complication of PID is an abscess, or collection of pus, in the ovaries, Fallopian tubes, and uterus. In some cases, this can become life threatening, requiring treatment with antibiotics or surgery.

To diagnose PID, a doctor begins with an internal pelvic exam. A woman with PID feels extreme pain when gentle pressure is placed on the cervix. After the exam, the doctor may order blood tests to check her white blood cell count. Most women with PID have elevated counts as a result of the infection. From here doctors may perform additional tests on the cervix. Scrapings and cultures taken from the cervix can be viewed under the microscope to help with the diagnosis. If the results are still not conclusive, doctors can look inside the abdomen of the patient after making a small incision and inserting a laparoscope, a fiber-optic viewing device, inside the body.

Not all bacterial PID infections are caused by sexual encounters. A few can be acquired during a vaginal delivery of a baby, a miscarriage, or an abortion. When the cause of PID is bacterial, antibiotics are used for treatment. Recurrence of the disease is common, but not necessarily dangerous. However, 40 percent of all women that have three or more episodes of PID become infertile. Sometimes the condition escalates to the point where the uterus must be removed.

Cervical Cancer

Viral STDs can contribute to the formation of cervical cancer in women. Cancer of the cervix is the most common cancer of the reproductive tract among pre-menopausal women. This cancer usually strikes women between the ages of thirty-five and fifty-five years of age. Most experts believe that exposure to the human papilloma virus, the cause of genital warts, contributes to the formation of cervical cancer. Some common symptoms of cervical cancer include abnormal bleeding between menstrual cycles, foul-smelling vaginal discharge, and discomfort during sex. When detected early, cervical cancer has a 90 percent cure rate. Sometimes, however, these symptoms do not materialize until the cancer has advanced to its late stages, allowing the cancer to spread throughout the body via the bloodstream and lymphatic system.

Cervical cancer usually grows very slowly, and yearly checkups can detect it in a manageable stage. Certain factors make some women more susceptible to cervical cancer. The early onset of sex (prior to twenty years old), multiple sex partners, history of genital warts, long-term use of contraceptives, and smoking add to the risk of developing this disease.

The treatment for this disease depends on the stage of the cancer when it is found. If the cancer is confined to the cervix, a surgeon may remove the cervix, but leave the uterus intact so the woman can still have children. If the woman is not planning to have children, the doctor may elect to remove both the cervix and the uterus in a procedure called a partial hysterectomy. Sometimes cells

Cervical cancer cells found in a woman's reproductive tract quickly multiply.

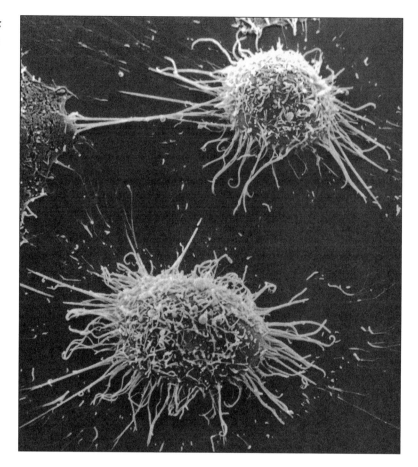

of cervical cancer penetrate beneath the surface of the cervix and enter small blood vessels and lymph nodes, permitting the cancer to spread to other parts of the body. If the cancer has passed beyond the cervix but is confined to the pelvic region, a total hysterectomy, removal of the uterus and ovaries, may be done. Radiation can be administered to kill any remaining cancer cells. Cancer that has spread beyond the pelvic region may be treated by chemotherapy, the use of anticancer drugs, or other medications.

Some Effects of STDs on Men

Although men lack a cervix, they can develop chronic diseases of the reproductive organs. Bacterial STDs can lead to the development of several inflammatory conditions. One of these is epididymitis, swelling of the epididymis, the tube that connects the testes to the prostate gland. A swollen epididymis causes pain, burning, and sometimes an abnormal discharge during urination. This condition can also result in low sperm counts. Early treatment with antibiotics can easily correct the disease. The most common causes of this disorder are chlamydial infection and gonorrhea. If the disorder is left untreated, the epididymis may be permanently scarred, resulting in infertility.

The prostate gland is also a common site of swelling and disease. Inflammation of the prostate, called prostatitis, is very common in men over fifty years old. Prostatitis can be the result of an STD, or it can occur if bacteria are transferred to the prostate gland from the colon or the bladder.

Prostatitis can be acute, occurring suddenly, or chronic, a long-term problem. Some symptoms include pain in the groin, chills, fever, low back pain, painful urination, painful ejaculation, and blood or pus in the urine. Diagnosis is based on symptoms and a physical exam. Doctors check the size and condition of the prostate gland by pressing against it during a rectal exam. Pressing also causes the prostate to release secretions that can be collected from the tip of the penis. The microorganisms that cause prostatitis are extremely hard to identify, but analysis of secretions in the laboratory may point to the culprit. If the offending organism is a bacterium, antibiotics can be given to treat the condition. Warm baths and aspirin are prescribed for infections that do not respond to antibiotics.

The prostate gland can also be the site of benign prostatic hyperplasia, a noncancerous condition that is very common in men over fifty years of age. In hyperplasia, the prostate enlarges, narrowing the width of the urethra and impairing the ability to urinate. As a result, urine flow decreases and the man has to strain to urinate. The bladder feels as if it is never empty, so the man urinates frequently. Because the volume and force of the urine stream declines, muscles of the bladder grow larger and stronger to force urine out of the body. When urine flow is impeded, infections can develop in the stagnant urine left in the bladder. The exact cause is not known, but scientists suspect it has something to do with hormonal changes that occur as men age. Diagnosis of this disorder is made in a fashion similar to prostatitis. Drugs may be given to relax the bladder, and sometimes a catheter must be used to remove excess urine from the bladder.

Prostate Cancer

An enlarged prostate does not increase a man's risk of prostate cancer. However, cancer can cause enlargement of the gland, so an enlarged prostate may be an indicator of a serious condition. Prostate cancer becomes more common as men age. Research indicates that fluctuating levels of hormones may play a role in prostate cancer. Cancer of the prostate gland is the most frequent cancer in men, with two hundred thousand new cases each year. It is the second leading cause of cancer death in men. Most prostate cancer grows very slowly, so if detected early, it can often be cured. It is remarkable to note that after surgeries and autopsies, prostate cancer was found in 50 percent of men over seventy and almost 100 percent of men over ninety years old.

The symptoms of prostate cancer are very similar to those of benign prostatic hyperplasia. The presence of a tumor in the prostate puts pressure on the urethra and interferes with normal urination. Pain and difficulty during urination are two common symptoms. Since early detection is so important and prostate cancer is very common, doctors encourage all men over fifty years old to have yearly exams of the prostate gland. During the exam a doctor checks for tumors or nodules on the prostate. If abnormal growths are found, further tests are ordered.

Blood tests and ultrasound procedures are two common follow-up tests used when suspicious nodules are

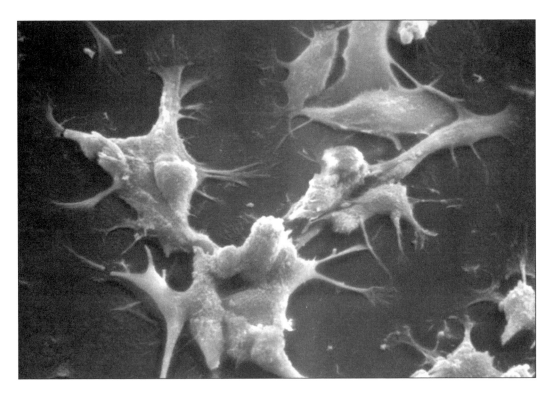

felt on the prostate gland. Blood is analyzed for the presence of prostate-specific antigen (PSA), which often is elevated in people with prostate cancer. A noninvasive test using ultrasound energy can give doctors a good picture of any abnormalities appearing on the prostate gland. Finally, a surgical biopsy of tissue from the prostate may be ordered. This tissue sample is examined under the microscope for cancer cells.

Prostrate cancer cells grow slowly in a culture dish.

The treatment of prostate cancer depends on the stage of the disease. Once diagnosed, further tests are done to see if the cancer has spread to other body locations. If the cancer is localized, radiation may be used to kill the cancer cells.

Surgery to remove all or part of the prostate is sometimes necessary. This surgery can cause the man to become impotent or incontinent. Doctors sometimes prescribe drugs to block the production of testosterone. This helps to slow the growth of tumors of the prostate.

The Final Word

Many, but not all, of the diseases that strike the reproductive organs of men and women are sexually transmitted diseases. Pathogenic organisms such as bacteria, viruses, and protists can cause these STDs. Abstinence, monogamous relationships, and safe sex can curtail the incidences of STDs in the population. Some pathogenic diseases of the reproductive tract have been linked to the development of cancer in the reproductive organs.

A gynecologist, a doctor specializing in female reproductive health, usually diagnoses reproductive disorders in women. Men may consult a general practitioner or another doctor, such as a urologist, a specialist of the urinary system, or a proctologist, who specializes in rectal disorders.

If not attended to quickly, some infections of the reproductive tract can lead to sterility in men and women. Cancers that strike the male and female reproductive organs are often manageable when diagnosed early. Early detection of diseases and disorders of the reproductive organs is made possible by getting yearly exams.

Medical Technologies of the Reproductive System

More than two hundred years ago, hospitals did not exist. Physicians visited patients in their homes, or opened offices where they could provide treatment. Sometimes, churches and monasteries provided long-term care for the very ill. However, in the 1800s, field hospitals were established to take care of men wounded during battle. Afterward, lots of them remained as centers for treating the sick. Within a few years, many expectant mothers chose to deliver their babies in hospitals where medical attention was available for the newborn. Early hospitals were very unsanitary places, however, and women who gave birth in those institutions often died of an infectious condition called puerperal fever.

An observant young Hungarian physician, Inaz P. Semmelweiss (1818–1865), was appalled when he found out that in any given year, almost one-fourth of the women on a maternity ward died. He launched an investigation to find out why. His inquiries revealed that most doctors worked in the morgue in the early mornings, and then went straight to the maternity wards without stopping to change clothes or even wash their hands. By issuing an order for all medical personnel to wash their hands and routinely scrub the maternity wards clean, Semmelweiss changed the way childbirth was managed forever.

Advances in medicine have taken many forms during history. Hand washing, a radical innovation in the 1800s, seems like common sense today. Because of modern technology and aseptic operating rooms, it is rare for women in industrialized nations to die during childbirth. In the last twenty years, modern medicine has introduced dozens of complex technologies that have enhanced the diagnosis and treatment of many conditions involving the human reproductive system. Some of the newest high-tech procedures help couples overcome reproductive problems that have prevented them from becoming parents. At the same time, science has perfected new sterilization techniques that prevent adults from having unplanned pregnancies.

The Very Useful Pap Test

One of the most important tasks performed by a gynecologist is a routine test called the Pap test. This proce-

A nurse helps a doctor into scrubs, sterilized clothing used to prevent the spread of infection.

dure, carried out during an exam of a woman's external and internal reproductive organs, helps doctors detect disorders such as cervical cancer. Named for its inventor, Dr. George Papanicolaou, the Pap test was introduced in 1928 to screen for abnormal cells of the cervix. However, it was not until the 1950s that the test gained widespread acceptance. During that decade the American Cancer Society (ACS) launched a campaign to teach American women the value of a Pap test.

The American Cancer Society recommends that the test be performed yearly on women who have had previous abnormal test findings, as well as women over forty years old. Since the 1950s the number of deaths due to cervical cancer has declined significantly. In fact, statistics show that of all cervical cancers diagnosed, about 80 to 85 percent were initially identified by inspection of Pap test results.

The Pap test can be performed by either a doctor or a trained nurse. During the procedure the patient lies on her back on the examining table. An instrument called a speculum is inserted into the patient's vagina to widen the vaginal canal, helping the doctor to get a good view of the cervix. A cotton swab moistened with a salty solution is inserted through the speculum and is gently rubbed over the cervix in order to extract some cervical cells. The cotton swab is removed from the woman's vagina and is then wiped across a microscope slide so that the cervical cells are transferred to the slide. The slide is dipped in a special solution to preserve the cells, and then sent to a medical laboratory for microscopic evaluation.

In the laboratory, a technician views the slide through a microscope and looks for any abnormal cells. Cells that are abnormal often have very large nuclei or other unusual characteristics. Once the microscopic evaluation is completed, lab personnel send a report to the doctor about the condition of the cervical cells. Cervical cells are characterized by placing them in one of five classes. A Class I report indicates that all cells appear to be normal.

A Class II reading means that some abnormal cells were seen, but they are not of the cancerous type. Sometimes infections such as *Trichomonas* or herpes can produce Class II, atypical cells. Cells that are described as Class III are atypical, and they may or may not be cancerous. A Class IV description indicates that cells are likely to be cancerous. Cells categorized in Class V are definitely cancerous. When any type of abnormal cells is found, a physician usually recommends that the patient have another Pap test or some other diagnostic procedure.

Pap Test Follow-Ups

In some cases, doctors follow up an abnormal Pap result with colposcopy. In this procedure, the doctor uses a magnifying lens to inspect the cervix for signs of cancer. This instrument (the colposcope), inserted in the vagina, magnifies the cervix about ten times its normal size. If suspicious cells are seen, the doctor may elect to remove a small amount of tissue from the cervix. This procedure, called an endocervical curettage, is performed by passing a small knifelike instrument through the length of the colposcope. Some cells are scraped from the suspicious area of the cervix and later examined under the microscope.

Colposcopy is one of the simplest techniques used to examine the reproductive tissues. A more invasive procedure, laparoscopy, can be performed to help a physician see inside the abdominal cavity. This test is useful in detecting endometriosis, ovarian cysts, uterine tumors, adhesions, pelvic inflammatory diseases, and even ectopic pregnancies. Prior to the laparoscopy, the patient is given local anesthesia, and then a small incision is made just below the navel. A thin fiber-optic telescope on the end of a hollow tube is passed through the incision and into the abdomen. Sometimes a gas such as carbon dioxide is injected into the abdomen so the abdominal wall is distended, giving the physician a better view. The doctor moves the laparoscope to the location inside the abdo-

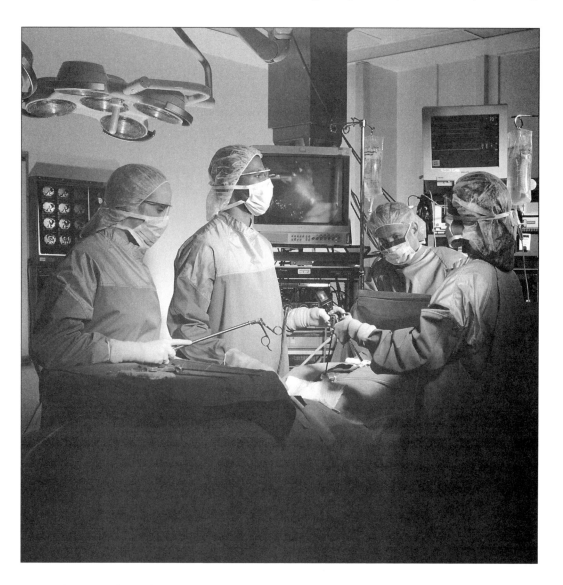

men that needs to be evaluated. If other instruments are needed, they can be passed into the abdomen through the opening of the laparoscope. Instruments can be used to biopsy suspicious tissue, remove foreign objects, or perform tubal ligation.

Doctors perform laparoscopic surgery to remove an ovarian cyst.

Another invasive examination procedure, a hysteroscopy, can be performed if a woman is suffering from unusual vaginal bleeding. First, a thin tube is passed through the cervix into the uterus. Then, the physician looks for the source of bleeding and inspects the uterus for any abnormal growths with the aid of a fiber-optic light inside the tube. If an abnormality is found, instruments can be passed through the hysteroscope to remove tissue samples for lab examination.

Sometimes this procedure is followed by a dilation and curettage (D and C). In this case, the patient is given general anesthesia, then the cervix is stretched open so that a curet, a curved knife, can be used to scrape the inside of the uterus. This technique can rid the uterus of abnormal endometrial tissue or unwanted material left behind after a miscarriage.

For Men Only

Men can also benefit from medical technologies of the reproductive tract. Prostate enlargement is a major problem for men. Benign prostate enlargement affects about 23 million American men, most over fifty years of age. When the prostate gland is enlarged, it often pinches shut a portion of the urethra. The result is that many men feel they cannot fully empty their bladder of urine, and for others urination is difficult and painful. Doctors often prescribe medications to relax the walls of the bladder and to shrink the size of the prostate gland. If the medications cannot relieve symptoms, a patient may elect to have a surgical procedure called transurethral resection of the prostate (TURP). This usually makes urination more comfortable, but 1 to 3 percent of men who have the surgery suffer from urinary incontinence afterward.

TURP is done while a thin endoscope (a laparoscopic tube containing a miniature camera) is passed through the opening of the penis. The endoscope is passed through the urethra and bladder to give the doctor a good view of the prostate gland and the urethra. An electric loop is threaded through the endoscope until it reaches the prostate gland. This loop is used to cut away small pieces of prostate tissue that are bulging into or blocking the urethra. Afterward, an electric current is sent through the loop to minimize bleeding. The trimmed pieces of the prostate gland are removed with the loop. These tissues are sent to a laboratory to be microscopically examined for the presence of any cancerous cells.

Passing the Pregnancy Test

Sophisticated medical technologies are also employed to assist in diagnosing, inducing, and evaluating pregnancy. One of the first tests that prospective mothers use to detect pregnancy is the home pregnancy test kit. The introduction of kits that can be used in the comfort of the home has saved many women who were not pregnant an unnecessary trip to the doctor. If a woman does have a positive reading in a home pregnancy test, she schedules a visit with her doctor to confirm the results.

A variety of home pregnancy tests are available to the public. Most of them function by detecting hormones called human chorionic gonadotropins (hCG). Levels of these hormones increase in a woman's body after an egg is fertilized. In most cases, they reach detectable concentrations in as little as ten days after conception. A negative result may mean that the woman is not pregnant, but it may be wise to repeat the procedure in a few days to make sure the test was not done too early.

Typical home pregnancy kits instruct the woman to collect a sample of urine first thing in the morning. A plastic-coated wand is then dipped into the urine sample. Most wands are covered with colored latex beads made of enzymes that are linked to antibodies of hCG. If enough hCG is present in the urine, the antibodies on the wand react with it and change color. If the hCG levels are very low, no color change occurs.

Technical Help

In the United States there are about 3 million infertile couples. Couples who have tried unsuccessfully for a year to have children are unlikely to conceive without help

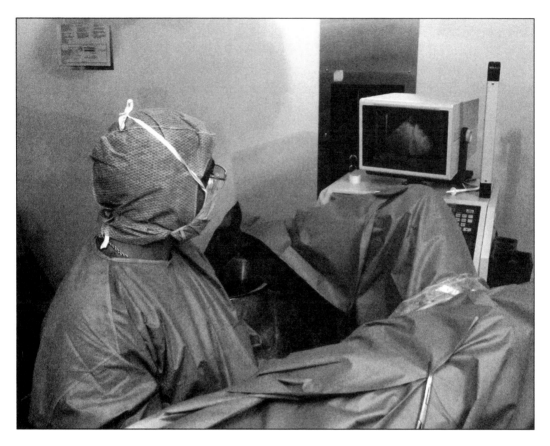

A doctor removes a woman's egg during a procedure that helps people who cannot conceive naturally.

from medical technology. The procedures available to such couples are called assisted reproductive technologies (ART). There are ARTs that help overcome both male and female reproductive challenges.

In vitro fertilization (IVF) is one type of ART. The first baby conceived by in vitro fertilization was born on July 25, 1978, in England. The birth of this "test tube baby" ushered in a new era of reproductive technology. Since that time, over five hundred thousand healthy babies have been born around the world as a result of IVF.

IVF is the treatment of choice for women who are infertile due to blockages in their Fallopian tubes. These women generally produce healthy eggs once a month, but the eggs cannot reach the Fallopian tubes where fertilization occurs. During IVF, mature eggs are removed from one of the woman's ovaries and then mixed with sperm cells in the laboratory. Sperm may be donated by the woman's husband or by another man. Medical technicians incubate the sperm and eggs for about forty hours, then observe them under a microscope. By that time fertilized eggs have undergone several mitotic divisions. Three or four fertilized eggs are injected back into the woman's uterus. If implantation occurs, the pregnancy then develops as any other pregnancy would.

The IVF Process

In most IVF procedures, doctors who are fertilization specialists administer medications to the woman to stimulate the ovaries prior to egg collection. The drugs induce the ovaries to release multiple eggs, rather than just one, during one ovulation cycle. Once the eggs are mature and ready to be harvested, the woman is given a sedative to help her relax. The doctor inserts an ultrasound probe into the vagina. A needle is passed through the probe. The probe helps the doctor to visualize the location of the ovary, and therefore helps guide the needle. Once the needle locates the eggs, several are aspirated or sucked into the needle. The probe and needle are then removed and the eggs are deposited in a lab container called a culture dish. The eggs are mixed with a special nutrient solution that is kept at a pH of 7.4 (slightly alkaline).

The sperm samples obtained from the donor by ejaculation are then introduced to the eggs in the culture dish. The sperm and eggs are incubated together at normal body temperature for about eighteen hours. The next day, laboratory technicians examine the contents of the culture dish under the microscope to determine whether any of the eggs have been fertilized and have started dividing. All fertilized eggs are then transferred to a second culture dish medium where they spend the next twenty-two to forty-six hours. In this time they divide to the eight- to sixteen-cell stage. Then they are ready to be transferred back to the woman's uterus.

Usually, two or three fertilized eggs will be placed inside the patient to improve the chances for a successful implantation. The fertilized eggs are transferred to the woman's cervix through a flexible tube that is inserted in the vagina. Afterward, progesterone is administered to promote the development of a thick uterine lining. Even though great care and exact procedures are followed during IVF, implantation is not guaranteed. The chance of successful implantation of fertilized eggs is only about 20 to 30 percent. However, if pregnancy does occur, it is not unusual for the IVF woman to give birth to multiple children. In fact 15 to 20 percent of all in vitro fertilization yields twins rather than a single baby.

After IVF

The mixing of eggs and sperm in culture media almost always produces more fertilized eggs than are needed for a single IVF. Many couples elect to freeze their unused fertilized eggs in case the first in vitro fertilization attempt fails, or in case they want to have more children in the future. With fertilized eggs available, the couple would not have to go through the process and expense of retrieving the eggs from the woman's ovary a second time.

To preserve unused fertilized eggs, specialists most often use a technique called embryo cryopreservation. This process involves the freezing of eggs in liquid nitrogen. Since this process was started in 1981, it has been a great success. Statistics show that approximately 75 percent of frozen genetic material survives the thawing and can be implanted successfully.

Couples considering cryopreservation often bring with them moral and ethical questions. Frozen fertilized eggs are not able to exercise legal rights, so there are no laws to determine what should happen to those that are not used by the parental couple. Presently, if a couple does not want to keep genetic materials created in IVF, the fertilized cells are either destroyed or donated to research.

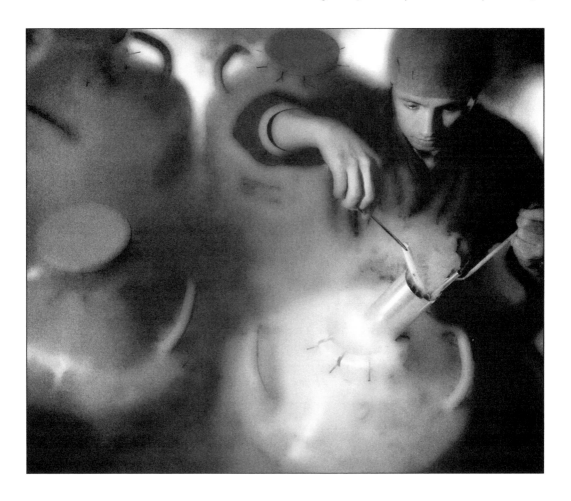

Already Pregnant

Once an embryo successfully implants in the uterus by either natural methods or by in vitro fertilization, a number of medical procedures are available to help ensure the health of the fetus. After pregnancy is established by medical tests, the expectant mother is often scheduled for a procedure called ultrasonography. This process uses high-frequency sound waves to locate and visualize the fetus or the expectant mother's internal organs. During the course of a pregnancy, an expectant mother may have several ultrasound examinations. Normally ultrasonography is performed very early in the pregnancy to

A technician works with cryopreserved embryos.

determine the number of fetuses present in the uterus, evaluate their size and condition, and predict the date of birth. Images taken later help obstetricians evaluate the development of the internal fetal organs and determine the sex of the baby.

Ultrasonography works much like the underwater sonar system used by ships. This process readily helps doctors see the condition of soft tissues that do not X-ray well. Prior to the procedure, the patient is often asked to drink a couple of quarts of fluid so the bladder will be full. A full bladder makes a good landmark during ultrasound. For the procedure the woman removes the clothing covering her abdomen and lies on her back on an examining table. The abdomen is covered with a lubricant such as mineral oil. The technician performing the procedure rubs a microphone-type device called a transducer over the abdomen. The transducer sends out sound waves and picks up the echoes they produce. These echoes travel to a computer that translates the sound waves into a picture on a monitor.

Ultrasonography can also be used to help guide instruments during two diagnostic tests, amniocentesis and chorionic villi sampling (CVS). Both of these tests can be used to detect genetic abnormalities in the developing fetus. In amniocentesis, a doctor inserts a sterile needle into the woman's abdomen and, guided by ultrasonographic images, gently pushes it into the clear region of the amniotic cavity. A small amount of fluid is withdrawn, with every attempt made not to injure the fetus. The fluid is sent to a lab where it is analyzed to identify any genetic disorders of the fetus. Amniocentesis is normally performed on women who are over thirty-four years of age or on women who have a history of genetic disease in their family.

Amniocentesis is performed after weeks fifteen to eighteen, since the amniotic fluid requires that long to develop properly. Analysis of amniotic fluid is useful because it contains fetal cells, as well as the urine pro-

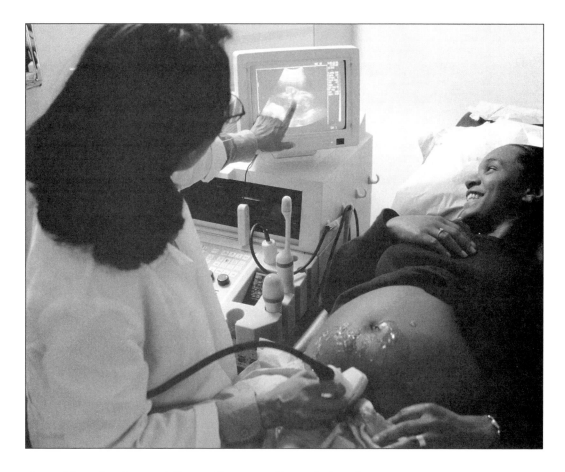

duced by the fetus. After cells are collected, it takes two to three weeks to get the results, since the sample must be cultured and analyzed in the lab. Even though the test is relatively safe, one out of every two hundred women experiences complications. For example, despite the use of ultrasound techniques to help the doctor guide the sampling needle, accidents can sometimes happen, perhaps leading to infection or miscarriage.

CVS can be performed earlier than amniocentesis. This procedure is similar to that of amniocentesis. However, instead of amniotic fluid, a sample of tissue containing chorionic villi is taken from the placenta. The advantages of this procedure over amniocentesis are that it can be done as early as nine to eleven weeks and the results of the

Specialists can monitor a baby's health and growth during an ultrasound.

test are received very quickly. Moreover, CVS does not require the puncturing of the abdominal wall. Rather, the doctor inserts a catheter into the vagina and through the cervix into the uterus where a tissue sample is removed. Complications occur about twice as frequently as with amniocentesis, however, and CVS does not determine as many fetal disorders as the more invasive procedure.

The Reproductive Wrap-Up

Humans, like all living things, reproduce to maintain their species. Like most multicellular organisms, humans engage in sexual reproduction, the method that assures the greatest genetic diversity among offspring.

In sexual reproduction, males and females produce different sex cells or gametes. Each gamete undergoes a series of cell divisions resulting in only half of the heritable material found in other human cells. When two sex cells fuse inside the female reproductive tract, they combine their half-sets of DNA to form a full complement of genetic information.

A zygote is a cell that results from the fusion of male and female gametes. The female reproductive tract provides a safe, nurturing environment for the zygote as it grows and develops. After about nine months, the once-tiny zygote is born as a fully developed human being.

Research performed by practitioners of modern medicine makes it possible to diagnose and treat many pathological conditions of the male and female reproductive systems. Some of the techniques used today are simplistic tests, while others are complex procedures. One simple but lifesaving procedure, the Pap test for women, helps doctors detect cervical cancer in the early stages when it is most able to be successfully treated. Other research has produced techniques such as TURP for treating conditions that affect the prostate gland.

In the past, infertile couples could never bear children of their own. However, today's technologies offer hope to childless couples. Procedures such as IVF and cryopreservation are options for those who cannot conceive a child without medical intervention. After a woman becomes pregnant, amazing technologies, such as ultrasound, amniocentesis, and CVS are employed to monitor the health and proper development of the unborn child.

GLOSSARY

biopsy: Removal of living tissue for clinical observation.

cesarean section: Surgical procedure used to remove a fetus from the mother when a vaginal delivery is impossible or a fetus is in distress.

chromosome: Threadlike structure in the nucleus of a cell that carries genetic information.

cleavage: An early embryonic phase consisting of rapid cell divisions.

conception: The fusing of a sperm and egg to produce a zygote.

ectopic pregnancy: Implantation of a fertilized egg outside the uterus.

egg: *See* ovum.

endometriosis: Abnormal release and growth of endometrial tissue outside of the uterus.

endometrium: The lining of the uterus.

endoscope: A lighted instrument that allows doctors to see inside the body.

fetus: In humans, an unborn child between the third month of development and birth.

gamete: A sex cell—that is, an egg or sperm.

gynecologist: A doctor specializing in diagnosis of disorders and administration of treatments of the female reproductive system.

hormones: Chemical messengers that regulate certain organs in the body.

lactation: The production and secretion of milk.

meiosis: A series of two cell divisions that cause the formation of gametes with one-half the number of chromosomes as a body cell.

menses: The monthly discharge from the uterus of a woman who is not pregnant.

mitosis: A single cell division that produces a clone of the parent cell.

ovulation: The release of an egg cell from the ovary.

ovum: A female gamete, or oocyte; informally, an egg cell.

placenta: Temporary organ that provides nourishment for an embryo and removes some wastes.

semen: A fluid produced by the male reproductive structures that contains sperm, nutrients, and mucus.

spermatozoan: A male gamete; informally, sperm.

tubal ligation: Sterilization procedure in females.

umbilical cord: A structure with arteries and veins that connects the placenta to the fetus.

zygote: A fertilized ovum.

FOR FURTHER READING

Elizabeth Fong, *Body Structures and Functions*. St. Louis, MO: Times Mirror/Mosby, 1987. Provides simple and thorough descriptions of various diseases of the human body.

Alma Guinness, *ABC's of the Human Body*. Pleasantville, NY: Reader's Digest Association, 1987. Discusses the various structures of the human body and addresses some interesting reasons for certain body functions.

The Handy Science Answer Book. Canton, MI: Visible Ink Press, 1997. Gives very cute explanations for a variety of happenings in the science world.

How in the World? Pleasantville, NY: Reader's Digest Association, 1990. This book provides interesting coverage of both physical and biological events that occur in life.

David E. Larson, *Mayo Clinic Family Health Book*. New York: William Morrow, 1996. Describes in simple terms the many diseases that can affect the human body.

Susan McKeever, *The Dorling Kindersley Science Encyclopedia*. New York: Dorling Kindersley, 1994. Gives concise information on physical and biological occurrences in life. Good illustrations help to explain topics.

Mary Lou Mulvihill, *Human Diseases*. Norwalk, CT: Appleton & Lange, 1995. Provides a good description of the most common diseases of the human body.

World Book Medical Encyclopedia. Chicago: World Book, 1995. Provides a vast amount of information on the physiology of human body systems.

WORKS CONSULTED

Books

Regina Avraham, *The Reproductive System*. Philadelphia: Chelsea House, 2001. Gives a good description of the male and female reproductive organs, the fertilization of the egg cell, and the development of the fetus.

Robert Berkow, *The Merck Manual of Medical Information*. New York: Pocket Books, 1997. Provides a detailed explanation of all organs. This book gives information on the causes, symptoms, diagnoses, and treatments of many diseases.

Charlotte Dienhart, *Basic Human Anatomy and Physiology*. Philadelphia: W.B. Saunders, 1979. This textbook covers the structure and function of all organ systems in the human body. It also provides information on symptoms and treatments of various diseases.

William C. Goldberg, *Clinical Physiology Made Ridiculously Simple*. Miami, FL: Med Masters, 1995. This booklet gives a very detailed explanation of body systems. Illustrations reinforce the written content.

John Hole Jr., *Essentials of Human Anatomy and Physiology*. Dubuque, IA: William C. Brown, 1992. This textbook of anatomy and physiology provides detailed explanations of the structure and function of all human body systems.

Anthony L. Komaroff, *Harvard Medical School Family Health Guide*. New York: Simon & Schuster, 1999. This book provides comprehensive coverage of the various disorders and diseases that can affect the human body. Symptoms, causes, diagnoses, and treatment options are provided.

Ann Kramer, *The Human Body, The World Book Encyclopedia of Science*. Chicago: World Book, 1987. Provides information on all body systems as well as giving explanations about unusual and interesting events that occur in the human body.

Stanley Loeb, *The Illustrated Guide to Diagnostic Tests*. Springhouse, PA: Springhouse Corporation, 1994. This medical book gives a very thorough description and explanation of how and why medical technologies are employed to diagnose and treat human diseases and disorders.

Robert Margotta, *The History of Medicine*. New York: Smithmark, 1996. This book gives a great synopsis of the history of medicine and how technologies have advanced over the centuries.

Elaine Marieb, *Human Anatomy and Physiology*. Redwood City, CA: Benjamin/Cummings, 1995. Offers a very detailed explanation of all human body structures and organs.

Steve Parker, *Medicine*. New York: Dorling Kindersley, 1995. This book explores the tools and techniques used in medicine to save lives over the past and present eras.

Websites

About (www.about.com). Easy-to-use site that offers information on all topics, including health and medicine.

About Children's Health (www.aboutchildrenshealth. com). Good information about all types of body systems.

Body Basics (www.kidshealth.org). This site presented by the Nemours Foundation is written for young people and explains the reproductive system in simple language.

CDC (www.cdc.gov). Information from the Centers for Disease Control and Prevention on any topic in health.

Children's Health (www.medem.com). Information on all types of children's health from the medical organization called Medem, Inc.

Cornell Medical College (www.edcenter.med. cornell.edu). The medical college of Cornell provides a wide range of information on body systems.

Countdown for Kids Magazine (www.jdf.org). Students can research any topics that interest them, including health and medicine.

11th Hour (www.blackwellscience.com). Valuable teacher resource for any type of information in science.

Fact Monster, Learning Network (www.factmonster. com). Provides information on all topics; suitable for any student. Provides a good science encyclopedia.

JAMA HIVAIDS Resource Center (www. ama–assn.org). The *Journal of the American Medical Association*, published by the American Medical Association, is a great resource for any topic in medicine.

MSN Search (www.search.msn.com). Provides a science library suitable for most students.

The Merck Manual (www.merck.com). This website gives a detailed explanation of body systems and diseases.

Women's Health Information Center (www. ama–assn.org). This site from the *Journal of the American Medical Association* covers a variety of topics that are of concern to women, such as reproductive issues and STDs.

Yucky Kids (www.nj.com). Easy-to-read articles on reproductive structures and other body systems.

Internet Sources

Dr. Seth G. Derman, M.D., "ART," 1997. www. members.aol.com.

"Assisted Reproductive Technologies Offer New Hope," University of Iowa Health Care, 2002. www.uihealthcare.com.

"How Designer Children Will Work," How Stuff Works, 2002. www.howstuffworks.com.

"How Human Reproduction Works," How Stuff Works, 2001. www.howstuffworks.com.

"How Prenatal Testing Works," How Stuff Works, 2001. www.howstuffworks.com.

"How the Reproductive System Works," Health Information Center, 2002. www.healthsquare.com.

"How Ultrasound Works," How Stuff Works, 2001. www.howstuffworks.com.

"Infectious Diseases," Methodist Health Care System, 2001. www.methodisthealth.com.

"Keeping the Reproductive System Healthy," American Medical Women's Association 1995. www.amwa-assn.org.

"Why Shouldn't Pregnant Women Smoke?" How Stuff Works, 2001. www.howstuffworks.com.

INDEX

American Cancer Society (ACS), 83

Aristotle, 33

asexual reproduction, 10–12, 19, 30

Braxton-Hicks contractions (false labor), 56–57, 59

breasts (mammary glands)
  development of, 43
  milk production and, 63
  structure of, 43–44

cells
  division of, 10
  fusion, 20
  interstitial, 23
  meiosis and, 17–18
  microscope and, 33
  mitosis and, 14–16
  oocytes, 34, 39
  parts of, 14

cervix
  cell classification, 83–84
  gonorrhea and, 70
  labor and, 61
  location of, 41
  mucus secreted by, 51
  pelvic inflammatory disease (PID) and, 74
  sperm and, 50

childbirth
  "afterbirth" and, 62

C-section (cesarean) and, 62
  delivery and, 61
  early understanding of, 33
  labor and, 59, 61
  midwives and, 65
  process of, 63
  reproductive organs involved in, 41
  sanitation and, 81–82
  stillbirth, 66
  vagina and, 41

chromosomes
  genetic material and, 11
  mitosis and, 14–16
  number of, in sperm, 25
  number of human, 13
  warp of, 19–20

circulatory system, 21
  development in embryo, 54
  pregnancy and, 58
  sexually transmitted diseases and, 70

circumcision, 29

clitoris, 35

Cowper's gland (bulbourethral), 29

digestive system, 21
  development of, 54

diseases
  bacterial, 69–72
  blindness and, 70
  cancer of penis, 68

PICTURE CREDITS

ABOUT THE AUTHORS

Both Pam Walker and Elaine Wood have degrees in biology and education from colleges in Georgia. They have taught science in grades seven through twelve since the mid–1980s.

Ms. Walker and Ms. Wood are coauthors of more than a dozen science teacher resource activity books and two science textbooks.